The Prison House of the Circuit

The Prison House of the Circuit

Politics of Control from Analog to Digital

Jeremy Packer, Paula Nuñez de Villavicencio, Alexander Monea, Kathleen Oswald, Kate Maddalena, and Joshua Reeves

MINNESOTA

University of Minnesota Press
Minneapolis
London

Portions of chapter 1 are adapted from "The Digital Body: Telegraph as Discourse Network" by Kate Maddalena and Jeremy Packer, *Theory, Culture, and Society* 32, no. 1 (2015): 93–117, https://doi.org/10.1177/0263276413520620, reprinted by permission. Portions of chapter 3 are adapted from "Police Media: The Governance of Territory, Speed, and Communication" by Joshua Reeves and Jeremy Packer, *Communication and Critical/Cultural Studies* 10, no. 4 (2013): 359–84, https://doi.org/10.1080/14791420.2013.835053, copyright National Communication Association, reprinted by permission of Taylor & Francis Ltd, http://www.tandfonline.com on behalf of National Communication Association. Portions of chapter 4 are adapted from "From Windscreen to Widescreen: Screening Technologies and Mobile Communication" by Jeremy Packer and Kathleen Oswald, *The Communication Review (Yverdon, Switzerland)* 13, no. 4 (2010): 309–39, https://doi.org/10.1080/10714421.2010.525478, reprinted by permission of Taylor & Francis Ltd, http://www.tandfonline.com. Portions of chapter 4 are also adapted from "Flow and Mobile Media: Broadcast Fixity to Digital Fluidity" by Kathleen Oswald and Jeremy Packer, in *Communication Matters: Materialist Approaches to Media, Mobility, and Networks*, Jeremy Packer and Stephen B. Crofts Wiley, eds. (Milton Park, Abingdon, Oxon; New York: Routledge, 2012), reprinted by permission of Taylor & Francis Group. Portions of chapter 6 are adapted from "Media Genealogy and the Politics of Archaeology: Introduction" by Alexander Monea and Jeremy Packer, *International Journal of Communication* 10 (2016): 3141–59.

Copyright 2023 by the Regents of the University of Minnesota

All rights reserved. No part of this publication may be reproduced, stored in a retrieval system, or transmitted, in any form or by any means, electronic, mechanical, photocopying, recording, or otherwise, without the prior written permission of the publisher.

Published by the University of Minnesota Press
111 Third Avenue South, Suite 290
Minneapolis, MN 55401-2520
http://www.upress.umn.edu

ISBN 978-1-5179-1416-5 (hc)
ISBN 978-1-5179-1417-2 (pb)

Library of Congress record available at https://lccn.loc.gov/2022039158

Printed in the United States of America on acid-free paper

The University of Minnesota is an equal-opportunity educator and employer.

UMP BmB 2023

Contents

Preface

About the Writing of This Book

This book has six authors—more than usual for most academic efforts in the humanities. We thought we should make a note about how it was written and why it was written that way. There are multiple versions of this explanation, and all of them are equally true, though any version without the others would be false. Some of the reasons for writing this book together were mundane, practical, and strategic, and others were more politically and theoretically intentional; as with all social phenomena that occur in space and time, it is difficult to parse whether "how" produced "why," or the other way around. And if you were to ask each of the authors individually, they might give you different answers. Do not ask us who we are, and don't ask us to remain the same; maybe we write together in order to have more than one face.[1]

While some of this book's chapters exist in different versions elsewhere, this work was always conceived as a book that would provide a coherent exemplification of the emerging methodology called media genealogy (see chapter 6 for the philosophical underpinnings of media genealogy as well as the subdisciplines from which it emerges and to which it contributes). In the early 2010s, Jeremy Packer met with a group of us, who were at the time North Carolina State University graduate students in the Communication, Rhetoric, and Digital Media PhD program, and proposed that we coauthor a set of media genealogy case studies as part of a multi-authored manuscript. While each chapter would initially have two coauthors, feedback and conceptual development was to occur via committee. It was to be a collectively authored set of discrete chapters. During the following years, others joined, careers developed, more research was conducted, and new thematic concerns emerged. Yet, with authors dispersed across North America, the completion of the manuscript called for a new mode. To this end, the coauthors met monthly via video chat during the spring and fall semesters of 2018 to draft a conceptual introduction and methodological overview, as well as to revise the existing chapters to better fit the coherent whole and a contemporary context.

We also wanted to include a workshop in the writing process—a face-to-face conversation with some of the contemporary voices that the volume cites as well as emerging media studies scholars at the University of Toronto. The workshop became the Media Genealogy Graduate Workshop, held May 17–19, 2019, at the University of Toronto Mississauga. The workshop is a perfect example of both noble and practical "whys": it gathered geographically dispersed colleagues around a table for a rewarding, co-present conversation. More pragmatically, it gave the authors a windfall of invaluable substantive feedback to guide manuscript revisions before a press and reviewers saw it. The formal respondents to the manuscript were Sheryl Hamilton (Carleton University), Colin Koopman (University of Oregon), Ganaele Langlois (York University), Liam Cole Young (Carleton University), and Greg Elmer (Ryerson University). The graduate students contributing to the workshop were Maya Hirschman, Brendan Smith, Bahar Nasirzadeh, Cansu Ekmekcioglu Dedeoglu, Michel Mersereau, Arun Jacob, Aakash Solanki, Alex Ross, Lee Wilkins, Bethany Berard, Stephan Struve, Jessica Chapman, and Evan Nevard. Chris Russill (Carleton University) also attended and provided commentary. The workshop provided a space in which we could *speak with* our ideal audience rather than simply trying to *write for* them. The people listed above, then, are authors with us in the most magnanimous (and, we think, truest) sense of the term.

An Introduction to the Circuit

Four Chairs, Four Problems

> *Allowing circulations to take place, of controlling them, sifting the good and the bad, ensuring that things are always in movement, constantly moving around, continually going from one point to another, but in such a way that the inherent dangers of this circulation are cancelled out.*
>
> Michel Foucault, *Security, Territory, Population*

A special code was devised: "Two taps of the bell, start dynamo; two additional taps, increase the pressure; one tap, stop."[1] In Auburn, New York, on August 6, 1890, this code was initiated in order to close a circuit connecting three humans with four technical objects: (1) a dynamo housed in a distant basement; (2) a 5- by 3.5-foot switchboard composed of an ammeter, a lamp board with thirty-six lamps, a regulating switch that governs the lamp board, and "the switch"; (3) a bell; and (4) a chair. A man throws "the switch," which rings the bell in a basement a thousand feet away. "Ding ding!" the code activates a technician who pulls a lever to turn on the dynamo. It purrs to life, sending 1,500 volts through the circuitry. Shortly after, the bell rings twice again, signaling for more electricity. Each signal prompts a human to increase the voltage. Surprisingly, it again rings twice, and soon thereafter two more sets of signals are sent requesting even more power. Later reports of the event describe the technician as confused:[2] "Just how much electricity is necessary?" Finally, the bell rings a single time, signaling that there is no longer any need for electricity. The dynamo is turned off. The circuit goes lifeless. William Kemmler, convicted murderer, also goes lifeless.

Strapped in the chair, Kemmler was the final element in this circuit, the last link in the first semi-automated electronic kill chain. A "doctor held a bright light to Kemmler's eyes, and the optic nerve showed no response."[3]

The autopsy noted that, among other physiological changes, "there was a decided change in the consistency and color of the brain."[4] His internal circuitry was converted to carbon; his capacity to collect, store, and process data had ended. To some, all had gone well in Auburn Prison that day. The four doctors who collectively performed the autopsy three hours after death all agreed that "unconsciousness was instantly produced and death was apparently painless."[5] Kemmler was the first human to be purposely drawn into an electrical circuit for the purpose of "humane"[6] execution, the eventual outcome of which had been a tenet in the debate over the standardization wars that pitted Edison against Westinghouse to determine AC or DC dominance.[7] Regardless of the victor, the daily work of the prison's generators was not execution, but rather powering the machinery of the prison factories. The execution required that the circulation of electricity be redirected to a single point, cutting into the electric-powered production of the prison.[8]

Outside the prison another hastily constructed circuit had been established by the Western Union Telegraph Company. The telegraph monopoly had built a temporary station within viewing proximity of the prison so that "dozens of telegraph operators"[9] could signal to the world that Kemmler was dead and that electricity had prevailed. So began the process of circuits signaling circuits about the triumph of circuits—an electric ouroboros.

In another kind of prison (a U.S. mental hospital in 1966), a second circuit-laden chair is devised—an electric chair of a different sort: an image-oriented behavior modification apparatus that is meant to cure sexual deviance. A young, attractive, verbal, intelligent, and successful[10] man who is attracted to other men decides that his attraction is a sickness and asks to be cured. His psychologist tells him about a procedure, and the gay man voluntarily straps himself into an apparatus. Electrodes brush his fingertips[11] while a screen projector shows him images of his lover that his doctor instructed him to bring to therapy, as well as images of men from gay magazines. The photos are associated with painful jolts of therapeutic electric energy. Eventually, images of naked women begin to flash across the screen, and these images act like a switch; they break the circuit and become associated with relief. Such aversion therapy, says the man's therapist, will "strengthen heterosexual feelings in the sexual response hierarchy,"[12] thus curing him of his homosexuality[13] and allowing him to walk free from social shame.

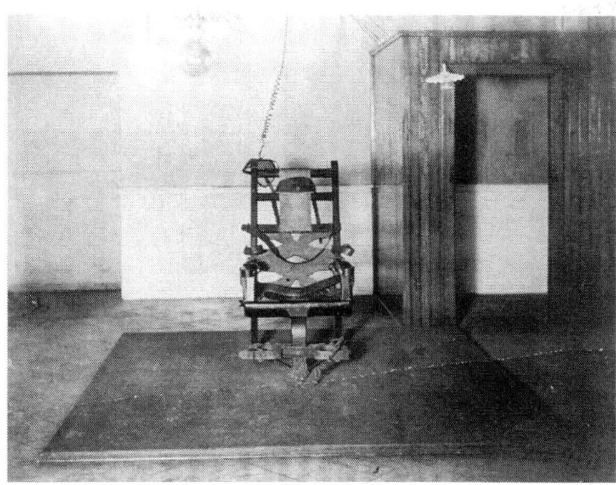

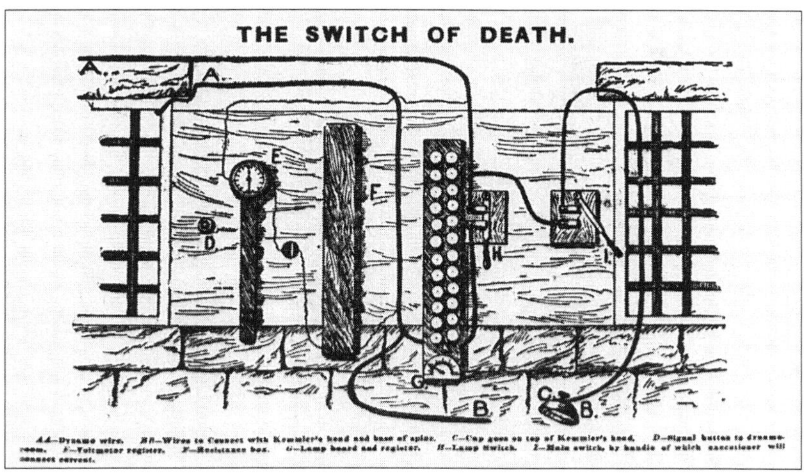

Figure I.1. Auburn Prison's electric chair, New York Auburn, ca. 1908. Photograph: https://www.loc.gov/item/2012646356/; "The Switch of Death," New York Evening World, *April 28, 1890.*

The electroshock gay conversion therapy circuit is not a final solution in the humanizing of execution, but rather a means to rewire the circuitry of the homosexual brain—desire redirected, behavior adjusted, and the brain purportedly changed. A different pathway for a different problem, but still wired to an institution. Arranging humans into different slots in a circuit can produce different effects. These slots were coded in simple binary terms, gay or normal. The human is not separate from the circuit, but an integral element of it. Yet, there are various roles that need to be played for a circuit to carry out its coding agenda.

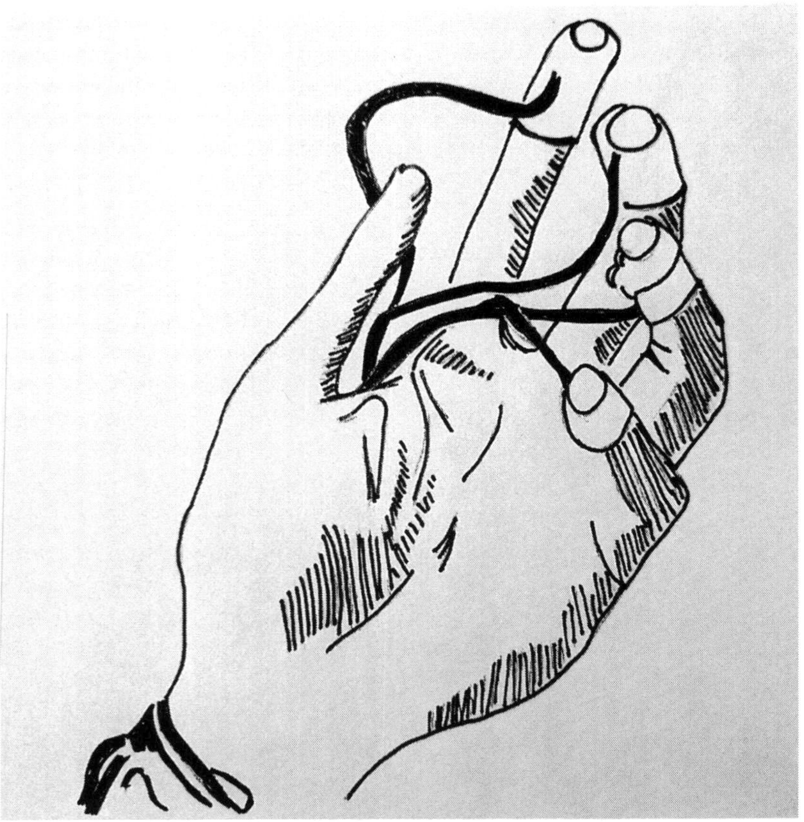

Figure I.2. Drawing of fingertip electrodes used in aversion therapy, as seen in "Gay Aversion Therapy 1970" by ABCLibrarySales. Created by Kate Maddalena.

In 2015, a third chair sits connected to a globally networked circuit that encircles the world, a circuit whose terminal nodes detonate in clusters of routine coordinates. The third chair can see and sense, as well as send very serious messages. As with our first chair, it too is connected to a kill switch—one that remotely executes. Our third chair involves another "switchboard," approximately 5 by 3.5 feet in size, which is lit by LED; the kill switch takes the form of a joystick. A keyboard sits to the left of the joystick, and a throttle, like the one in a plane cockpit, flanks the other side. This mundane apparatus links to a wall of hardware, including a telephone and a series of monitors that visualize a collection of data sets: altitude, geographical coordinates, speed, fuel usage, time stamps, and real-time video footage beamed in from electro-optical infrared cameras flying thousands of miles away. On a few of the screens, moving images scan a monotonous pixelated territory. Occasionally, the scan slows, stops, and homes in on a particular site. Reminiscent of our second chair, a blander sort of pornography fills these screens: a fuzzy pair of figures chatting outside a house, a man smoking a cigarette at the base of a hill, a few cars traveling down a mountainside road. Suddenly the phone rings. Code received. Target confirmed. The throttle is thrust slightly downward, then upward again. A red button atop the joystick is pressed, and a new circuit is enlivened. As the cargo is released, this new circuit routes between a satellite in outer space and the pixelated man at the bottom of the hill. As the payload rushes toward the target to complete the circuit, the pilot dutifully watches from his chair. Suddenly a flash of white fills the screen, as dirt and shrubs and smoke fly in all directions. When the smoke clears, our pilot watches the man at the bottom of the hill writhe on the ground, his dark silhouette from the infrared camera slowly fading to white. The new circuit's mission—its only mission—is terminated after only a few seconds. After standing for a moment and wiping his face, our pilot returns to the chair and again pulls the throttle.

Figure I.3. Drone Throne. U.S. Air Force photograph by Airman First Class Michael.

A fourth chair could have sat empty, seeing as the labor of its occupant had been fully automated by a complex set of circuitries. The chair itself features three different motors that allow it to comfortably cradle its passenger. Beyond the immediacy of the chair, thirty other motors dictate the micro and macro movements of the vehicle. Forward-facing radar, GPS, eight cameras, and sonar provide raw data for the Level 2 automated system's computational algorithms that made sure there was no need for a human to be seated in this chair. However, Joshua Brown remained seated even as his Tesla's autopilot mode sent his car directly into and under the path of a bright white eighteen-wheeled truck that, to human perception, clearly intersected its path. The bottom section of the Tesla made its way under the truck and continued on for a short distance as its electric motor and brain hummed along. Joshua died at the scene, killed by his car's inability to separate signal (white eighteen-wheeler) from noise (bright sunlight). In the days following Brown's accidental execution, Elon Musk would attempt to redirect blame to another media system for potential deaths related to automotive automation. "If, in writing some article that's negative, you effectively dissuade people from using an autonomous vehicle," he told reporters, "you're killing people."[14] Failure to capitulate to the circuit was tantamount to death. Technophobic leftovers from journalism's old-media regime could be counted on for at least one last round of fear mongering, misdirecting their ire onto the apparent cure. Years after his physical death, Brown's ill-fated and sadly ironic YouTube post, "Tesla Autopilot v7.0 Stop and Go Traffic," continues to extol the virtues of his Tesla's autopilot. Brown described the feeling of acquiescing to the circuitry in this way: "Now you don't have to worry about anything. Just let it go."

These four chairs give insight into the very material ways in which circuits interpenetrate human bodies, how they shape human experience in the name of solving problems. Each is enabled by unique codes. Each connects technical apparatuses, governmental initiatives, and human bodies. The carceral space and the space of execution are just two such interlocked elements in the circuitry. One decides who sits in the electric chair, quite possibly a circuit judge who makes the rounds from court to court as a form of juridical circuitry that maintains continuity, while another circuit provides the necessary electrical power to permanently short-circuit the brain's fragile signaling network. Humans at times are mere conduits in very complex machines whose powers they are completely unable to resist. Power flows through them, but they serve small parts in immensely

Figure I.4. Screenshot from YouTube, "Tesla Autopilot v7.0 Stop and Go Traffic," by Joshua Brown.

complex socio-technical circuitry. In other instances, humans are actuators. They press buttons or pull switches in order to keep an already-begun process pushing forward. In some rare cases, they design and build the very circuits that mold and shape the behaviors and movements of human and nonhuman physical/biological/social processes. But if we stand up from the chair and look around, disconnecting ourselves from our own circuitry, where do we stand? What can we see? Is anyone or anything capable of escaping the circuit?

Circuitry and Circulations

The circuit is a technical mode of governance based on managing, producing, monitoring, stopping, and activating flows, often through a process aimed at managing risk and securing predictable outcomes. The broadest understandings of the circuit intersect both spatial and temporal dimensions. On the one hand, a circuit can produce a fixed spatial order by referring to a "line, route, or movement that starts and finishes at the same place."[15] On the other, we see the inception of a temporal regularity for how space will be occupied and enlivened: "an established itinerary of events or venues used for a particular activity."[16] Circuits are more than logistical media.[17] They reconfigure existing media through the addition of information layers that enact increasingly specific and predictive patterns of movement, behavior, and consciousness through a process of encircuiting. The circuit is more productively conceived as the application of screening technologies that modulate the flow of information (including bodies) through space and across time. Whereas logistical media normalize and make efficient, the logic of the circuit is to specify and make effective. And while circuits can be understood to be metaphorical, it is much more important to recognize them as simultaneously physical, conceptual, and logical. They circumscribe and canalize physical and conceptual space and coordinate temporal flows through these spaces in accordance with particular governmental logics. Circuits exist in time and space, and they operate according to logics of coding and capturing that enact differing modes of governance, where "to govern" is understood as "the right disposition of things, arranged so as to lead to a convenient end."[18]

Circuits often manifest in infrastructures of circulation, in what Matthew Tiessen and Greg Elmer call "the pre-conditions of conditions."[19] We can see the centrality of circulation as good government in the most rudimentary yet life-sustaining infrastructural circuits, such as the waterworks of a major city like Singapore.[20] As Mark Usher notes regarding the centrality of this problematic to the work of governance, "governmentality was originally directed towards circulation and the 'material instruments' through which it flowed, from the widening of roads to the navigability of canals, constructed to provision the town and strengthen the power of the state."[21] For the past several hundred years, international relations have been conceptualized according to the logics and demands of circulation.[22] While clean water and adequate sewage at the local level may convince us to think of such circuits in a positive light, efficiency is not always unquestionably good. An efficient, shared water system can also enable a community to efficiently share a disease.[23] However, beyond the unintended effects, water infrastructure is more than a logistics of efficiency; it also has a circuit logic of specifying who gets water, how much, when and where, and of what quality. The circulation of clean and dirty water affects new and intensifies preexisting power relations. It determines where different people can live hygienically at what cost and for how long. The circulation of water necessarily reorganizes the circulation of bodies, the circulation of capital, and the circulation of inequity. Historically ghost towns might result from the drying up of a well or a river changing course. Flint, Michigan, a predominantly Black and working-class city, stands as a symbol of a contemporary ghost town in which poisoned drinking water was purposely allowed to circulate through its waterworks to the considerable detriment of its population. In terms of governance, all problematics can be turned into a binary equation: How should things circulate or not circulate?[24]

Foucault and the Prison House: From Soul to Circuit

Kemmler's execution in the electric chair was no accident. And we chose it not by accident but in reference to the gruesome opening of Foucault's *Discipline and Punish: The Birth of the Prison,* which describes a botched attempt to draw and quarter Robert-François Damiens, a failed kingslayer. Foucault's text immediately follows this description with the minutiae of a prison timetable created eighty years after Damiens's execution. For Foucault, the timetable and execution are representative of two vastly different penal modes: one rooted in a spectacular display of the sovereign's power, and another that had its roots in the humanistic enterprise of reforming subjects so as to remake them into productive members of society. The latter disciplinary objectives exemplified in the prison spread to other institutions and produced a robust disciplinary society whose various techniques for monitoring, shaping, and directing its citizenry spread across the social realm and became foundational to the modern state and the rise of capitalism. Further, "the birth of the prison" referenced both the rise of the penitentiary and the establishment of the soul as a kind of self-disciplined prison for the body (inverting Plato's notion that the body was the prison house of the soul). Foucault summarized it as such: "The soul, effect and instrument of a political anatomy; the soul, prison of the body."[25] We want to suggest that something else has taken form that works across both the social realm of the distribution of power/knowledge relations and the realm of the ethical processes of subjectification. We suggest that new forms of technical media and systems of networked distribution, both based in the logic of the circuit, have wildly reworked the terrain of governance, security, self, and subject. The governmental logic that produces the circuit coexisted with the logic that produced the soul, but our new reliance on the materials that manage electrical flows has produced a new epoch. The circuit is to the circuit board as block is to city, as station is to nation, as server is to farm.

We draw heavily from Foucault, who identified key elements in this reorientation: "At any rate, under this name of a society of security, I would like simply to investigate whether there really is a general economy

a

b

Figure I.5. Examples of circuits to illustrate the range of the circuit's governance: microchip, city, transportation system, market, and brain. (a) *Circuit board. Copyright Raimond Spekking / CC BY-SA 4.0 (via Wikimedia Commons).* (b) *Screenshot of a city in Google Maps.* (c) *Sydney Trains Map. Created by MDRX / CC BY-SA 4.0 (via Wikimedia Commons).* (d) *Advertisement from Gigamon showing a server rack.* (e) *Brain as a circuit network.*

Berowra
Mount Kuring-gai
Mount Colah
Asquith
Hornsby
Waitara
Wahroonga
Warrawee
Turramurra
Pymble
Gordon
Killara
Lindfield
Roseville
Chatswood
Artarmon
St Leonards
Wollstonecraft
Waverton
North Sydney
Milsons Point
Circular Quay
Wynyard
Martin Place
Town Hall
St James
Kings Cross
Edgecliff
Bondi Junction
Museum
Central
Redfern
Normanhurst
Thornleigh
Pennant Hills
Beecroft
Cheltenham
Epping
Eastwood
Denistone
West Ryde
Meadowbank
Rhodes
Concord West
North Strathfield
Richmond
East Richmond
Clarendon
Windsor
Mulgrave
Vineyard
Riverstone
Schofields
Quakers Hill
Marayong
Emu Plains
Penrith
Kingswood
Werrington
St Marys
Mount Druitt
Rooty Hill
Doonside
Blacktown
Seven Hills
Toongabbie
Pendle Hill
Wentworthville
Westmead
Parramatta
Harris Park
Granville
Clyde
Auburn
Lidcombe
Olympic Park
Flemington
Homebush
Strathfield
Burwood
Croydon
Ashfield
Summer Hill
Lewisham
Petersham
Stanmore
Newtown
Macdonaldtown
Erskineville
St Peters
Sydenham
Tempe
Green Square
Mascot
Domestic Airport
International Airport
Wolli Creek
Arncliffe
Banksia
Rockdale
Kogarah
Carlton
Allawah
Hurstville
Penshurst
Mortdale
Oatley
Como
Jannali
Sutherland
Kirrawee
Gymea
Miranda
Caringbah
Woolooware
Cronulla
Loftus
Engadine
Heathcote
Waterfall
Merrylands
Guildford
Yennora
Fairfield
Canley Vale
Cabramatta
Warwick Farm
Liverpool
Casula
Glenfield
Leppington
Edmondson Park
Carramar
Villawood
Leightonfield
Chester Hill
Sefton
Birrong
Yagoona
Bankstown
Punchbowl
Wiley Park
Lakemba
Belmore
Campsie
Canterbury
Hurlstone Park
Dulwich Hill
Marrickville
Turrella
Bardwell Park
Bexley North
Kingsgrove
Beverly Hills
Narwee
Riverwood
Padstow
Revesby
Panania
East Hills
Holsworthy
Macquarie Fields
Ingleburn
Minto
Leumeah
Campbelltown
Macarthur

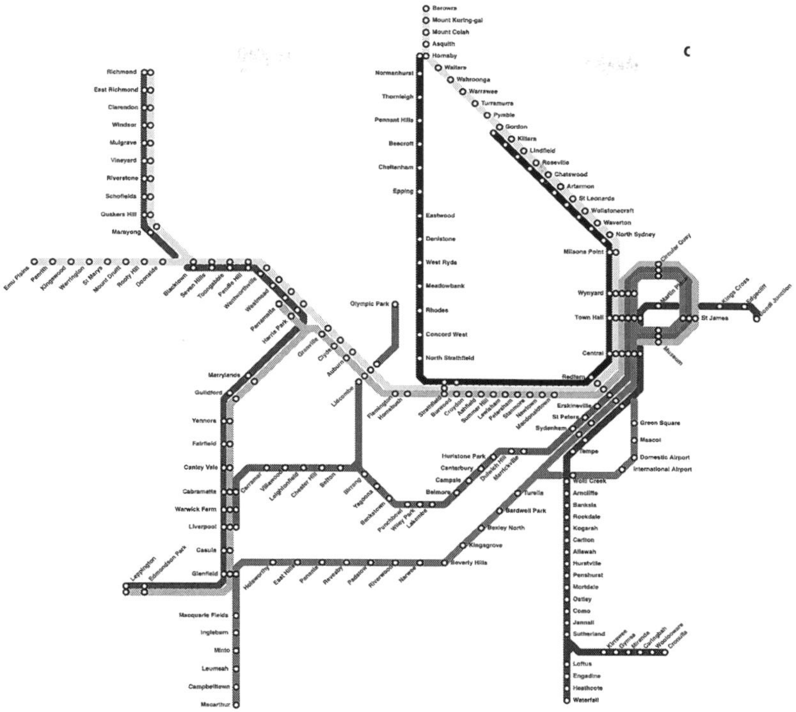

c

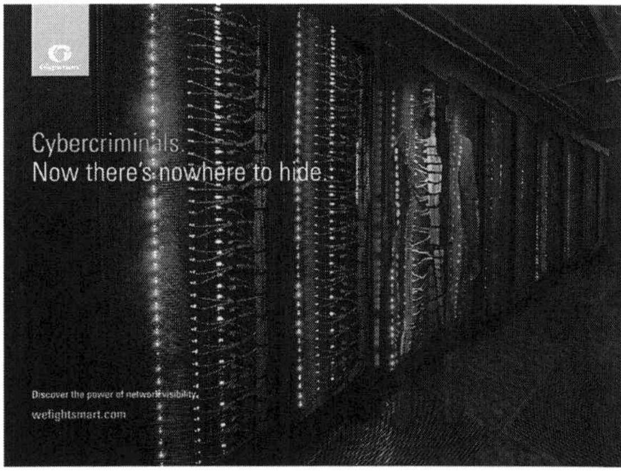

Cybercriminals.
Now there's nowhere to hide.
Discover the power of network visibility.
wefightsmart.com

d

e

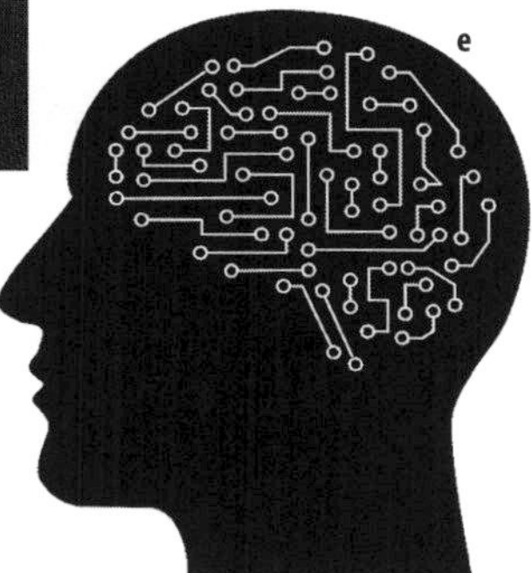

of power which has the form [of], or which is at any rate dominated by, the technology of security."[26] Mark Salter highlights this dual process taking place in Foucault's work in his volume *Making Things International 1: Circuits and Motion*. According to Salter, from an international perspective, "security and sovereignty as modes of governance are intimately connected with the management of circulation, and in our case particularly international circulation. . . . To understand sovereignty, security, and circulation, we must understand how things and people move, and how they are made to move in which particular circuits."[27] In such an analysis the circulation of things, technologies, and nonhuman practices takes on greater importance of analysis for understanding international sovereign relationships than they were previously given. We wholeheartedly agree, but we want to suggest, as does Sean Cubitt in *The Practice of Light*, that the management of circuits, at the macro-level *and* micro(chip)-level is equally a part of the process of managing circulations, providing expressions of governance, and forming relations of power. As Cubitt suggests, the demand for statistical optimization that Foucault explained was fundamental to governing through circulation, "govern[ing] not only power and money but also the minute and constant operation of visual media."[28] There is no sense in which we can think of the circulation of people, populations, intercontinental ballistic missiles, currencies, and ocean tankers without understanding that they are guided, monitored, or stood in for by the circulation of electrical currents and pulses of light. Circulations have burrowed into increasingly smaller circuits and time frames even while they expand ever further and across longer durations. Our suggestion is that Foucault's genealogical analysis, an analysis that moves from logics of disciplinarity to logics of security and circulation, is dependent upon "the circuit" as a set of social, epistemological, political, and technical practices that are enlivened by power, knowledge, and subjectification.

As such, this book investigates various logics of the circuit in their material forms as a means of understanding some of the ways that the prison house of the soul, as described by Foucault in *Discipline and Punish*, has largely been replaced by what we call *the prison house of the circuit*. While it is true that circulation and mobility were part of the Foucauldian vocabulary and conceptual framework,[29] for the most part Foucault, for better or worse, has too often been thought of as a theorist of confinement.[30] Yet shifts from a society organized along disciplinary logics of confinement, fixed spaces, tightly managed timetables, normalized and fixed identities, hierarchical knowledge production and dissemination, and attempts to eradicate undesirable phenomena have been reoriented toward (1) flexible, multidimensional, extrahuman, and distributed epistemologies; (2) managed distribution of risk and monitored experimentation with freedom and radical openness; and (3) repurposing of human–machine interactions that deemphasizes fixed individual identities through the creation of networked, real-time, algorithmic modes of self-assessment, ethical intertwinement, and delegated agency. While Gilles Deleuze has famously described some of these changes as being representative of a shift to a "control society," we want to elucidate and augment the key considerations that Deleuze identifies with this shift, while more specifically focusing upon historical transition points during which such shifts are evidenced as well as the media and media practices that enable them.

When we describe what media and media practices are, we follow McLuhan, Foucault, and others by using the term *technology*. Foucault focuses on *tekhne*, "a practical rationality governed by a conscious goal."[31] Recognizing the possible conflation with the term *technology* understood too narrowly in his view as "hard technology, the technology of wood, of fire, of electricity," Foucault asserts we also should broaden our understanding of technology. He claims, for instance, that "government is also a function of technology: the government of individuals, the government of souls, the government of the self by the self, the government of families, the government of children, and so on."[32] Thus the historical study of a specific *tekhne* would demand situating it within the broader sense of the word to understand how it functions in terms of a practical rationality and a means to govern specific practices and peoples. Such an approach could lead us to ask: Why and in what ways have encircuiting and encircuited media been imagined as an arena of analysis and application that could accomplish such practical goals?

The first full-scale elaboration of the work of Foucault as it relates to how he utilized the term *technology* appears in Michael Behrent's article "Foucault and Technology."[33] The article's very focused analysis aptly draws out how much Foucault depended upon the term as he developed his theory of power beginning in the early 1970s. For Behrent, Foucault's interest in power does not merely overlap with his renewed technological vocabulary and his development of a genealogical method. Rather they emerge as interlinking concepts. When we write "technology," we necessarily connote "technologies of power."

Foucault's use of the term *technology* is then fundamentally associated with two key considerations of genealogy as they pertain to the media archaeological tradition. First, Foucault's antihumanism or posthumanism called for a new language or set of terms that could describe the mechanisms by which human beings were worked upon in such a way that did not imply a destruction or alteration of their human essence. Technologies of power, power technologies, or techniques of power provide this terminology and suggest that the same types of knowledge and techniques used to shape nonhuman elements and abstractions were also used on human bodies. Second, his turn toward genealogy, the analysis of power, becomes an analysis of specific techniques of power. It is not a broad analysis of the roots of power. It is not a search for the origins of power, but rather a very specific analysis of the technologies of power that come to enliven or give life to the modern state and modern institutions, even the modern military. Discipline, as merely one example, was imagined by Jeremy Bentham as a kind of "universal machine" that could shape what the human might become in a similar sense that Alan Turing imagined the computer could be adapted to perform nearly any technological function. Other specific techniques include technologies of the self, technologies of knowledge production, and technologies of governance (which produce an immanent governmentality).

This last set of techniques, the government of self and others, is now a thoroughly technical, data-rich, and experimental set of circulatory entwinements. The establishment of circuitry as a foundational practice demands incessant experimentation with the material and conceptual world that reflexively calls forth its own dissolution and reboot: a governmental feedback loop of media escalation. Friedrich Kittler goes so far as to argue that circuits encode humans into themselves. And, as the circuitry becomes increasingly complex, humans are necessarily penetrated by circuits in expanding ways, humans function as actuators:

> What remains of people is what media can store and communicate. What counts are not the messages or the content with which they equip so-called souls for the duration of a technological era, but rather (and in strict accordance with McLuhan) their circuits, the very schematism of perceptibility.[34]

Rather than humans writing code as autonomous actors in circuits, humans must eventually be overwritten by circuits. The chapters that follow act as points of entry for understanding this process. They do not tell the whole story, but they are exemplary of how the encoding takes place. In the remainder of this Introduction we would like to outline how we are conceptualizing the term *circuit* and how it differs from other popular theoretical frameworks for examining the power relations inherent in technologically mediated governance.

Conceptualizing the Circuit

Circuits are understood to manage circulations or flows, to make them possible. Natural flows are guided by evolution and the laws of physics and chemistry; mechanical flows by efficiency and entropy; political and social flows by security, discipline, and similar forms of power; monetary flows by risk analysis and class struggle. Typically, a circuit is a physical path, consisting of one or more wires, roads, tracks, lanes, zones, passages, tubes, pipes, or some other mechanism that contains and directs a flow. Circuits tend to integrate intermediate switching points, Ys, splitters, and other mechanisms of separation, conjoining, divergence, and feedback. Circuits are protected, secured, sped up, rearranged, and attacked. While circuits are infrastructural, in their more metaphorical and theoretical existence they cannot be reduced to infrastructure. Think, for instance, of how an annual influenza epidemic is managed in terms of its circulation.[35] Numerous existing "infrastructures" may be drawn into the logistical formulation of public-health efforts, but influenza is not merely an infrastructural problem; it must be problematized in terms of circulation. In the mobilities literature, scholarship calls for a "shift towards a broadened account of circulatory security" that pays attention not merely to nodes but to the open processes of circulation in its entirety, as with a pipeline.[36] As more of the world and its various forms of physical, geological, social, and biological properties are made knowable through circuits (through the so-called internet of things, but clearly already begun during the Closed World[37] era of the Cold War), what was once a process of discrete circuits bearing specific materials to and through a set of nodes begins to morph into an open or "smooth" space[38] in which time and space themselves are co-produced by circuitry: the map is the territory;[39] the circuit is the Earth.

Armand Mattelart argued that circulation is the backbone upon which modern governance and the modern nation-state came into being.[40] He provided a shorthand for drawing together a vast set of interrelated epistemological, political, and social projects under the rubric of the term *communication*. In his classic work on the "invention" of communication, Mattelart considers communication "from a wider viewpoint, encompassing the multiple circuits of exchange and circulation of goods, people, and messages."[41] For Mattelart, circuits and circulation—communication—came to be the defining feature that drew together natural physiological understandings of networks (such as the circulatory system) with the establishment of various political philosophies—including political economy via "political anatomy" and the physiocrats—that were founded on promoting and orchestrating "natural" circuits that would guide and monitor the flow of populations, money, data, propaganda, news, disease, and even war. For the period Mattelart is most interested in, the body and other "natural" phenomena are employed in the logical organization of systems of governance. The period we investigate works in reverse. Technical logics and metaphors are increasingly used as a means for understanding the human or the natural world more broadly.[42]

Foucault also emphasized the primacy of circulation to the political and social imagination of eighteenth- and nineteenth-century Western Europe. As the populations of Western Europe became more mobile in the nineteenth century, governing authorities and their allies among the merchant class began carrying out surveillance on beggars, vagrants, criminals, and other suspect people passing through urban areas. The form of government developed to address this new population flow, according to Foucault, was at heart "a matter of organizing circulation, eliminating its dangerous elements, making a division between good and bad circulation, and maximizing the good circulation by diminishing the bad."[43] It was, in other words, a matter of classifying goods, ideas, and people as worthy of circulation or unworthy of circulation—as Foucault puts it, it strives to build "an intensity of circulations: circulation of ideas, of wills, and of orders, and also commercial circulation."[44] Rousseau imagined Nantes, the city, like a heart: "We can see that the problem was circulation, that is to say, for the town to be a perfect agent of circulation it had to have the form of a heart that ensures the circulation of blood."[45] Those things from whom positive social value can be extracted should be allowed to circulate; those who might taint the social bloodstream should be identified, neutralized, and removed from circulation, or, in a compromise, redirected into circuits of discipline or security that might protect the public from being contaminated by them.

While English usage of the term *circuit* dates to late Middle English, its etymological roots are in Old French by way of Latin. The Latin *circuitus* was a variant of *circumire* ("go round"), from *circum*, meaning "around," and *ire*, meaning "go." The French term *circuire* hosts a cluster of connotations and is synonymous with to fool, trick, take in, deceive, and ensnare. A second, related set of meanings suggests orbiting, surrounding, and encircling. We may be drawn into a circuit's orbit, surrounded by nothing but the unfolding of its tricky logic. The circuit may not only govern; it may fool or deceive so as to ensnare. If you have ever accidentally completed an electric circuit with your body—an electric fence or a poorly grounded lamp plug—you understand the implications of a circuit with no switch. The power flows through you while simultaneously immobilizing you, changing the beat of your heart for (if you're lucky) just a moment. The switch is accessible to the circuit designer and the circuit user, but the man in the electric chair can't touch it. If he could, the switch would allow access to escape. In most cases, though, control of the switch is direct control of the circulation of power.

In terms of today's media, the electrical circuit is the building block of electronic computation, AI, and autonomous machinery. Most simply, a circuit is a single path, a closed loop, on which electricity can travel. More important, electrical circuits are the fundamental building blocks for all electronic and computational media. An integrated set of circuits, sometimes called a chip, microchip, or IC, is a collection of electronic circuits on a semiconducting material, often silicon. The silicon chip is simply a grouping of miniaturized circuits. Electronic computation is produced by governing the circulation of electricity within these miniature circuits. Circuits are mechanisms that draw numerous elements together in order to rationally and repeatedly produce a dynamic range of results. Circuits depend upon signals, codes, and switch points. Circuits produce discrete paths to organize and monitor flows. More broadly, a contemporary digital network is (merely) an arrangement of circuits. According to Kittler, software, or code, doesn't exist as such. "All code operations, despite their metaphoric faculties such as 'call' or 'return,' come down to absolutely local string manipulations and that is, I am afraid, to signifiers of voltage differences."[46] It's all hardware. It's circuits all the way down.

With the concept of the circuit, we offer a complementary counterpoint to the clean, discrete, secret-keeping, and boundary-maintaining frame of digitally enabled control. If the digital and its related epistemologies of control (cybernetics, cryptography, etc.) are focused on producing security by keeping people out, the circuit functions to produce security by drawing people in. No mode of encryption can ever depend upon a key that permanently exists outside itself. It must, in the end, by virtue of its mating, be circuitous. Media historians and theorists[47] have fleshed out related conceptions of and around digital media,[48] and their use of the term typically accesses a history in which the management of circulation strives to exclude certain subjects and include others. This story is embedded in the history of computing and directly linked to World War II's Allied code-breaking (discourse-oriented) projects.[49] Our story of the circuit is embedded in the same history but is linked instead to those militaries' projects in the governance of bodies and their productive potential (biopolitically oriented).

In our conceptualization, circuits constitute an emergent mode of power relations that is distinct from pastoral, sovereign, and disciplinary power in that they depend upon new media of governance and security that are fundamentally constructed by and through circulation. It is the circuit's aim to produce a perfect state of security by removing human noise from governmental communication channels. Chapter 3 describes predictive policing in such terms. Nonetheless, the circuit depends on human bodies as part of its technical apparatus, so it must render those bodies docile, which it does by proliferating media and discourse that can be captured and used experimentally to refine and expand the circuit. There may be different modes of circuiting, some of which "silence" human noise, some of which "amplify" human input in order to understand it more fully, and some of which "modulate/attenuate" human input in order to make the circuit work efficiently.

In electrical circuits, a path of conductive materials must be created in order for electricity to flow and have an effect on the world. In order for electrical power to circulate, a connection must be made; a relationship must be forged. Circuitous logics have been operational throughout history across different regimes of power, yet they function most seamlessly in the cloud, where other logics lose ground, water, and air. Benjamin Bratton explains that The Stack is an "accidental megastructure" that "smooths space by striating it with heavy physical grids of cables and server farms, and striates space by smoothing it out with ubiquitous access, sensing, relay, and processing micropoints."[50] In this open field of ubiquitous digital access to the spaces, behavior, and patterns of movement and consumption, circuit logics increasingly enable and constrain movement, identity, tastes, and possibilities. They open us to the possibilities of data-driven and algorithmically processed modes of power.

The flow of electricity naturally moves from higher voltage (the "power source") to elements in the circuit with lower voltage. Yet circuits and their various elements are able to manipulate, split, and store electrical current in such a way that the directionality and the force of electricity can create different effects and affect a wide range of interrelated objects. In other words, circuits fine-tune the brute power of electricity and shape its effectivity. The productive capacity of electricity is thereby enhanced even as its immediate voltage is diminished and spread thin. The behavior of electrons in an electrical circuit is very similar to the establishment of power relations through the technologies of governance that cultural studies scholars discuss in terms of security. Carrying this analogy through, we might think of sovereignty as sometimes acting like lightning: an unwieldy destructive force that lashes out in ways that are not carefully measured or equally distributed. Such power achieves results through a spectacular show of fire, light, sound, and violence. Technologies of governance such as security, governmentality, and biopolitics work through circuits that move information, bodies, and materials in highly governed and monitored fashion so as to build out relations of capillary power that are light, fluid, and dispersed. Lightness, however, should not be confused with weakness or impotence.

Technically, circuit switching is the process by which two nodes are brought together to create a fully dedicated connection between previously disconnected elements of a network of circuits. Conversely, circuit unswitching is akin to disconnecting a circuit or at least disarticulating two segments of a network. Switches, then, are the consciously designed and enacted means for making circuits and for having control over which segments are encircuited, for how long, and at what times. Such control may also imply the capacity to use circuit switching to enact a "kill switch" that disables the entirety of the circuit. Unswitching a connection between two nodes is not in and of itself a "kill switch," except in cases with no redundancies, or where only two nodes exist, or where one circuit is entirely dependent upon another circuit. For instance, the electric chair circuit is both enabled and disabled through a switch that connects it to the electricity circuit. In chapter 2 of this volume, two separate circuits—the circulation of triaged soldiers and the circulation of medical records—were brought together through the "medical field card." The field card functions as a form of circuit switch: two nodes from two different networks combine to form robust biosubjects that circulate within a national biopolitical circuit.

Ensnarement, or what we are calling "encircuiting," is the circuit's particular mode of subject production/use of the body. It is the telos of the circuit. We can isolate at least three modes of encircuiting, though in any historical context encircuiting might occur in multiple or hybrid forms across the three (or more) modes of encircuiting we have isolated here. First, a subject may consciously choose to become encircuited, voluntarily submitting all or some part of itself to the circuit's security-producing apparatus. This is usually done in exchange for certain social or personal benefits. This can clearly be seen in the case of automobility, where registering for a driver's license and entering into the circulation of traffic might be considered optional despite the inherent difficulties of living without automobility. In exchange for becoming encircuited in a complex system of laws, policing, insurance, car service and inspections, tolls, parking, and so forth, the driver is offered greatly enhanced mobility and supports the maintenance of critical infrastructures of transportation that empower economic, political, and cultural production. Similarly, in the case of smart glasses the human consciously enters the information circuit due to a desire to improve the self. They stay within the circuit when the so-called improvement transitions to the new norm of visual information consumption, production, and storage, and our previous visual state becomes a deficit.

Already implicit in this first mode of encircuiting is the second, where an encircuited subjectivity is positioned as the default and new subjects become encircuited due to the logic of normativity. The police circuit has historically been most often activated by civilian calls and is powered largely by citizen surveillance, where the normalization of turning to and interacting with police helps power the circuit.[51] This is often combined with surreptitious police surveillance—wiretaps, pen registers, tailing, staking out locations, crime scene forensics, cell-phone geolocation and stingray technologies, and social-media monitoring. Becoming encircuited here triggers mandatory forms of encircuitment, like the collection of fingerprints and biometric data and the establishment of rap sheets, that serve to further encircuit criminalized subjects. That said, it is worth noting that police circuits are not equitably distributed; certain communities and locales find themselves ensnared in circuits of intensified—and more frequent—police scrutiny. While we might all be encircuited within the police circuit, the effects and severities of that encircuiting are differential in all the predictable ways.

The last form of encircuiting we have isolated is a fully mandatory form. This is most easily seen in the example of forced conscription and the Selective Service Act in the United States in particular. Encircuiting here is mandated by the state with harsh penalties for refusal and strong mechanisms for enforcing that mandate. Regardless of the mechanism of encircuiting, the resulting circuits share a common feature that Colin Koopman has described as "fastening," where our bodies and the information through which we understand ourselves and navigate the world become fastened to data regimes operated by powerful structures like the state, international corporations, or large technical infrastructures.[52] This fastening makes it exceedingly difficult to exit a circuit after having been encircuited. More often, encircuited subjects stay within these circuits, modifying themselves or being modified such that they become compatible with the circuit, continually producing information that can be extracted by the circuit to further ensnare them and enhance its own power.

While circuits are often imagined and rhetorically positioned as omnipotent, when met with the messiness of the material world in their actual deployment they can behave unexpectedly, break down, and be co-opted. Two of the most prominent versions of this can be understood as short circuits and open circuits. Short circuits are failures of the circuit due to circuit overload. They are not designed, but they may be either accidental or intentional. Short circuits result from the unmitigated flow of power through a circuit that is unable to accommodate the load. The daily commute's traffic jam as discussed in chapter 4 is a simple yet highly visible example. Circuits overload, and elements within them burn out and produce a break in the circuit. We can describe a short circuit as the overapplication of power into a circuit. The relationship established between elements in the circuit don't match the demands from one side or the other. For example, in chapter 5 we note that the technological augmentation of vision can become overloaded, resulting in madness, incoherence, or disorientation for the person completing a visual information circuit.

The open circuit is the ur-problem of the circuit. An open circuit exists where the connection (or relationship) between the power source and an element of the circuit is broken, exited, or disconnected. In simple terms, an open circuit is one that no longer works because connection is lost. Open circuits are often caused by the shortcomings of the human body and forces of nature. And they are, of course, a problem often solved by ameliorative technologies. In chapter 1 we describe the clumsy, nascent circuitry of what will later become the precise machine of the communication technology of modern warfare. This circuit was often inadequate to the demands of its charge. To close the open circuit of the soldier's often-broken line of sight, for example, the Signal Corps invents "chromosemic rockets" deployed by a signal pistol that encoded messages high in the air.

What the Circuit Is Not, or an Overview of the Book

One of the key cases we make throughout the book is that the elaboration of a new concept of the circuit bears critical and analytical fruit. It adds to contemporary media studies discourses on the digital, computation, technical media, infrastructure, flows, networks, platforms, logistics, and control by highlighting undertheorized aspects of contemporary technoscientific processes. One can best grasp the unique dimensions and historical trajectory of the circuit by engaging a method that we call media genealogy, which is fully elaborated in chapter 6. There we draw on the work of Foucault and his interlocutors to articulate a fuller theory and method of media genealogy, demonstrating how it offers a more politicized and diachronic analysis of how circuits are invented, implemented, stabilized, and interconnected, as well as how they are adapted over time. We will show that circuits come to perform an essential function in producing the "grid of possibility," governing who we can be and what we can think, know, sense, and intuit. By drawing on historical scholarship inflected by Foucault, ranging from cultural studies to German media studies to Anglo-American and French theories of science and technology studies, we can get a clearer articulation of exactly what the circuit *is* and what it is *not*.

The circuit is not only digital. Digital media are discontinuous, non-semantic, and modular media for the sake of constituting large amounts of information and for taking apart and remaking that information to interpret and deploy it. These media are a material set of commitments to the process of abstraction, discrete categorization,[53] and modularity that allow for optimal data collection, storage, and processing by knowledge-making systems. These commitments and their concurrent media practices also function as ways of being, making, and knowing. The circuit may employ the digital as an enervating logic (and indeed, the logic of the digital is perfect for the circuit's purposes, since it renders a discourse with no semantic slippage), but in those cases the digital is part of the cultural context[54] of the circuit. The digital worldview[55] obscures any view beyond the circuit. The two terms are by no means congruent. Where the digital parses, producing objects by atomistic means,[56] the circuit totalizes, modulating the flow of objects (digital or otherwise) through time and space. The circuit's apparatus connects the digital and the nondigital by either delimiting or ensnaring the semantic power of the subject.

Chapter 1 takes up the genealogical problem of the semantic subject in an emergent media apparatus bent on the subject's erasure. In it we provide an example of digitality and discipline wherein the human body becomes receptive to doing communicative work in which it is not a meaning-making, politically significant element. We consider the foundation of the U.S. Signal Corps during the Civil War a moment at which the soldier becomes ensnared in a proto-postdiscursive communication circuit. We treat the Signal Corps's practice of flag telegraphy (1) historically, as a manifestation of the ideas of one man to solve more than one problem; then (2) genealogically, as an apparatus that produces a certain kind of soldier–subject; and then (3) in terms of media apparatus, the various mediations and remediations that happen between networked materials in order to "get the message through" (the Signal Corps's motto). Finally, we step back to look at these treatments as a history of the present, tying the contemporary logic of circuitry to an apparatus that the Civil War–era Signal Corps first enabled and reified. The genealogical approach, we argue, allows for a view of the subject in the circuit that is at once more nuanced and more terrifying than the viewpoint of straight Kittlerian archaeology.

The circuit is not a network. The network is perhaps the most infrequently defined term to be near ubiquitously employed in media studies scholarship. The concept spills across disciplines, inspiring equally Gilles Deleuze and Félix Guattari's anticapitalist and anti-state notion of the rhizome and the type of DARPA-funded cybernetics defense research that W. H. Auden described as being "conceived in the hot womb of violence."[57] Today, the term seems most often to be used to describe an arrangement of entities (nodes) and the relations between them (edges) that can orchestrate the communication of information or data asynchronously and nonhierarchically. These forms of networks were originally designed to escape the limitations of serial processing that had been standardized in the integrated circuit of solid-state electronics, already imagined by John von Neumann as his hardware arrays were being standardized.[58] The most famous iteration of this form of network is perhaps Paul Baran's "hot potato heuristic routing doctrine," which overcame the limitations of hardware by outsourcing the intelligence to route and switch communications to both peripheral nodes in a network as well as to the messages themselves.[59] Donald Davies further developed this technology and dubbed it packet switching, forming a backbone for later TCP/IP protocols and the internet writ large.[60] As George Dyson writes, "All our networking protocols . . . are simply a way of allowing hundreds of millions of individual processors to tune selectively to each other's signals, free of interference, as they wish."[61] The end result is a network topology that appears random, designed more by nature than by humans, in which complex results obtain from simple rules.

There are certain limitations to understanding networks and circulation more broadly solely through decentralized packet switching, most notably its obfuscation of the way power continues to operate in these networks structuring the grids of intelligibility and possibility that constrain human and machine actions. As Alexander Galloway and Eugene Thacker explain, "*The juncture between sovereignty and networks* is the place where the apparent contradictions in which we live can best be understood. . . . Networks, by their mere existence, are not liberating; they exercise novel forms of control."[62] They note that while no one individual controls a network, networks do maintain power relations and are just as likely to be organized centrally (with pyramidal or hierarchical control schemes) or decentrally (with control radiating out in a hub-and-spokes model) as they are to be organized in a distributed fashion (as in the packet-switching model).[63] Chapter 2 will demonstrate how the concept of the circuit can more clearly capture not only these variations in network models but also their hybridity. Circuits can be arranged serially or in parallel, individually, in a network, or in an internetwork. Each circuit can be centralized, decentralized, or distributed. We examine the first global and real-time communication, command, and control circuits that were used to automate military logistics in World War I and establish the necessary linkages to allow for the circulation of bodies and health records associated with those bodies. Here we look to the use of military medicine, statistics, and anthropometry in the mobilization for and conduct of combat in World War I to see how soldiers' bodies were circulated for maximum efficiency. These circuits bear a closer resemblance to the original networks developed by J. C. R. Licklider and Robert W. Taylor, who envisioned a close coupling of human and machine in a symbiotic partnership where each maximized the strengths of the other—the speed of computation met the goal-setting intelligence of the human.[64] Here the circuit channels information to decision makers and simultaneously communicates their commands, orchestrating the war effort and collaboratively managing troop strength and maintaining the domestic war economy. Here we see tight circuits of automation and human direction interlinked into multiple network forms. One key example of this is triage, which operated in decentralized and even distributed modes at the front while simultaneously circulating information back to a centralized statistical office of the U.S. Surgeon General of the Army to be analyzed in digested form at the national level and used to assess the ongoing strength of the military.

The circuit is not control. On the one hand, control appeared as a mere "footnote" or "postscript" to Deleuze's philosophy. Yet on the other hand, control is something of a microcosm or "realization" of Deleuze's entire ontology.[65] We have become, especially since Galloway's *Protocol,* accustomed to thinking of Deleuze as a philosopher of media and a prophet of the digital era. While in one sense that is obviously true, media technology remains on the very periphery of Deleuze's oeuvre.[66] Control, the Deleuzean theory that is most often taken up in media studies, is little more than a five-page afterthought. Despite this lack of original detail, however, control has proven to be one of the most generative concepts in contemporary media and cultural theory.

Control derives its conceptual significance from a historical claim—that "control societies" are now following upon Foucault's legendary sovereign and disciplinary societies. Control's key term, *modulation*, is the essential activity of this emergent epoch. In Deleuze's words, controls are distinct from the "enclosures" and "molds" of disciplinary societies because they are "a modulation, like a self-deforming cast that will continuously change from one moment to the other, or like a sieve whose mesh will transmute from point to point."[67] The metaphor, here, is textural—subjects move through ever-modulating casts that form and deform according to the pressures of "a continuous control," a "perpetual training."[68] And this textural metaphor is especially well suited to the analysis of subjects and subject formation.

Yet subject formation is not the primary object of the circuit; it is *incidental* to it. While control might be an effect of circulation, the logic of circuitry exceeds the subject and, hence, exceeds the historical specificity embedded in the sovereignty/discipline/control triangle. If a control modulates, the circuit circulates. It circulates subjects, to be sure, but it eschews the textural constraints of subject formation in favor of the slippery, intuitive absorption akin to transduction. The textural friction of replicative modulation is thus only, at most, its by-product, and the patient training process characteristic of control is replaced by a persistent, unrelenting circulation, the purpose of which is to erode the friction of the objects it processes, evades, or absorbs. The human body, accordingly, is not the main object or target of circulation. It is at times its agent and at times its victim. But it is always, ultimately, its impediment.

Accordingly, control is frequently positioned as a form of power that no longer bothers to constrain individuation or subjectification and instead focuses on modulating sub-individual or "dividual" flows in accordance with capitalist ends. Galloway and Thacker note this explicitly and argue that control can "operate at a level that is anonymous and non-human, which is to say material," preferring not to intervene in individuation but instead to modulate its outputs at the dividual level.[69] Antoinette Rouvroy makes a similar argument in the development of her concept of "algorithmic governmentality." For Rouvroy, control requires the digital surveillance of behavior to build user profiles, which are then used to modulate the dividual. This process no longer produces subjects, but instead is a process of "de-subjectification" in which individuals are fragmented and unable to maintain singular, coherent identities and social groups are only formed based on statistical abstractions.[70] As Deleuze notes, we move from the use of distinct molds as enclosures that structure the formation of subjects to the modulation of "auto-deforming" molds that continuously change across time and space and from person to person.[71]

Hence it is not so much that control refrains from enclosing space as it is that control moves from manipulating Euclidean to non-Euclidean space. Control is about manipulating topology, such that apparently unenclosed spaces can be invisibly deformed to delimit the possibilities for movement or action in accordance with the ends of power. Our freedom to individuate ourselves is an illusion, always anticipated and modulated by regulatory systems that deform the space of possibility at increasingly precise points in space and time. Control initiates butterfly effects, making microscoping, dividual interventions that cascade into global impacts. In chapter 3 we analyze this through one of the more fitting examples of circulation inscribed through multiple circuits of power: policing. The professionalization of policing at the turn of the twentieth century amid rapid urbanization and population growth offers unique insights, as limited resources required the interconnection of a biopolitical circuit (statistical management of population-level criminality) with fewer and fewer disciplinary circuits (concrete interactions with embodied police officers and criminal justice institutions). This arrangement was only made tenable by the addition of control circuits that instituted forms of self-policing, community surveillance, and restructured (non-Euclidean) urban space to minimize the spatial and temporal distance of all points on the urban grid from the police. In its emphasis on the use of circulation to produce law-abiding citizens, this chapter demonstrates how circuits of control can be inscribed within larger circulations to modulate dividual behavior while still performing subject formation and the management of social groups in accordance with collective norms like criminality. This is a helpful corrective to the articulation of control by media theorists focused on more recent case studies, demonstrating that the abandonment of subjectivation and individuation is not essential to the operation of control but instead a historically contingent aspect of contemporary platform-based surveillance capitalism. This chapter also helps to demonstrate the genealogy of control and inscribe it within longer histories of circuit-based power formations. So much of the emphasis in media studies scholarship on control gets placed on digital computation, but this is perhaps a product of translation. As Yuk Hui notes, Deleuze's use of the term *numérique* has a double meaning; it can just as easily be used to refer to digital technologies as it can to refer to the use of numbers for statistical management.[72] Chapter 3 explores this slippage in the term, demonstrating how circuits of control emerged earlier in the analog context of policing and modulating criminality.

The circuit is not equivalent to infrastructure. Infrastructures are often positioned as large-scale, vertically integrated, monopolistic, long duration, ubiquitous systems that utilize standards and gateways to network together heterogeneous systems.[73] They are often examined as large technical systems, like electric power grids, telephone networks, air traffic control systems, or international internet service provision.[74] While these phenomena might be understood as well through the logic of the circuit—Paul Edwards in fact describes such technological systems as "the connective tissues and the circulatory systems of modernity"[75]—in our understanding the circuit is a more flexible concept. The circuit is more likely to privilege canalization of space and time, probabilistic intervention, forking and feedback, and augmentation or modulation of preexisting materials and forms than the monopolistic imposition of standards at large scale through enduring, vertically integrated, and ubiquitous systems. While both infrastructure and the circuit produce subjects through a process of learned membership, the circuit does not necessarily ensnare people through the typically infrastructural qualities of ubiquity, reliability, and durability.[76]

Chapter 4 takes up the case of the circuit logic of automobility. While circuits constitute a large part of the infrastructure of automobility—from the race circuit to the dozens of CPUs required to run a modern automobile—understanding the circuit *as infrastructure* misses the larger point that the circuit constitutes the primary logic by which mobility is thought. When we consider the circuit alongside infrastructure, it leads us to many interesting projects, such as the incorporation of CPUs to meet regulatory requirements, the relationship between mobility and entertainment through onboard entertainment systems, or a close look at the development of car-based cellular telephony and the role of circuitous infrastructure in making that possible. We acknowledge the role of the circuit *as* infrastructure without reducing it *to* infrastructure.

Beginning with an exploration of Raymond Williams's *mobile privatization* as a problematization of communication and mobility in the postwar era, we track a move from public technologies to privatized ones in the automobile circuit. We ultimately suggest that *screening technologies*—a host of technologies that leverage storage, connectivity, and mobility in order to increase capacities for control, informationalization, and convergence—are reorienting the automobile circuit from a private experience of personal ownership to a personalized experience managed by data brokers in a shared (but importantly not public) model. In conceptualizing what may otherwise be considered "logistical media" as screening technologies, we are able to productively describe a shift in the governing logic of automobility from one of rigid disciplining and canalization to a more flexible modulation of encircuited people, places, and objects that constitutes a new field automated management at a distance. In other words, in focusing on the circuit we are less concerned with the management of flows via logistics than the rapid development of an increasingly granular field upon which screening technologies can intervene. This examination of the automobile circuit also reveals a shift from navigation between fixed sites such as home, work, and school to an always connected access to digital spaces. Importantly, the reverse is also true: digital spaces also have access to us as we participate in these systems via screens. Understanding the ways in which mobility is reconfigured by and through technologies of screening gestures toward newly configured circuits of mobility and of the public, private, and personal.

Finally, the circuit is not a platform, although a platform can be connected to/in a circuit. As Tarleton Gillespie notes, a platform is a slippery concept. It operates as a "sociotechnical assemblage" and a "complex institution," combining the wills and ends of many in a complicated flow of actions and content. Gillespie argues that platforms are online sites or services that: "a) host, organize, and circulate users' shared content or social interactions for them, b) without having produced or commissioned (the bulk of) that content, c) built on an infrastructure, beneath that circulation of information, for processing data for customer service, advertising, and profit . . . [and d)] moderate the content and activity of users, using some logistics of detection, review, and enforcement."[77] As José van Dijck argues, this means that platforms constitute an ever shifting governance of communication balancing users' desire for self-expression with the economic and legal logic of platform owners seeking to channel users' self-expression toward profit making.[78] Similarly, Ganaele Langlois and Greg Elmer argue that platforms are fundamentally structured by economic interests that predefine the communicative acts that are possible on and through the platform in accordance with the platform owners' profit-seeking logics.[79]

At this point we can clearly see that platforms are much more likely to operate exclusively in the commercial sector and focus on achieving capitalist economic ends. We can thus at first glance see a vague difference between them and circuits, which can operate just as easily in the state sector and toward governmental ends. Perhaps, then, platforms are a contemporary subset of circuits. Toward that end, it is worth piecing together how they govern circulation. As Carliss Baldwin and Jason Woodward explain, platforms operate modularly, with a set of core components that remain relatively stable and govern the interactions of modules and the mechanisms through which third-party developers and users interface with the system. This architecture allows for complementary components to constantly be developed, both by the platform owner and by third-party developers and users, allowing content, features, and functions to proliferate atop a stable architecture.[80] Jean-Christophe Plantin, Carl Lagoze, Paul N. Edwards, and Christian Sandvig explain that platforms can thus be viewed as forgoing vertical integration in favor of seeking extension by third parties and users that follow the platform's core rules, growing as it mediates the distribution of content made by others without having to pay them. This offers benefits to users, who gain a standardized interface for all the platform's content and functions, and to developers, who have a prefabricated code base, large audience, and easy access to marketing information. The platform is able to monetize its intermediation between developers and users through network effects, as both users and developers become increasingly locked into the platform as it grows.[81] And what's more, this structure allows platforms to position themselves as neutral intermediaries with little control over the content they circulate, thus obscuring the ways in which they shape that content and user behavior in accordance with their own economic and legal ends.[82] As Nick Srnicek aptly summarizes, "Platforms, in sum, are a new type of firm; they are characterised by providing the infrastructure to intermediate between different user groups, by displaying monopoly tendencies driven by network effects, by employing cross-subsidisation to draw in different user groups, and by having a designed core architecture that governs the interaction possibilities. . . . They are an extractive apparatus for data."[83]

Platforms are thus born-digital, internet-centered content aggregators and disseminators that work to create "walled gardens" on the internet. As Srnicek notes, their logical end is to fragment the internet.[84] This is because the network effects that drive their profit making require monopolies on data extraction. Circuits thus differ from platforms in a number of key ways. While the platform requires cloud computation and digital code to manage its modular expansion, circuits can just as easily operate in the realm of the analog. One can clearly see the antagonism between the two concepts as platforms work to extend themselves offline, capturing more of the real world, producing virtual worlds, and augmenting reality, all of which requires them to negotiate with the analog, with noncommercial entities, and with human embodiment, and all of which has been done for nearly a century by circuits, as demonstrated in previous chapters. Chapter 5 explores how the circuit, rather than the platform, is a more useful concept to explore this overlapping of digital and analog worlds.

We examine the long history of the technological enhancement of vision—from analog spectacles of the Renaissance to smart glasses of the present day—to normalize, through the management of visual data, expectations in areas of labor production and sociality. Considering the fact that sight has come to dominate humancentric understandings of reality and scientific rationality[85] and that vision is a series of mechanical and biological processing of photons, the examination of light processing to create the human world becomes a media-specific problem to be solved by circuits.[86]

We are primarily concerned with how modes of visual problematization always already assume a scientifically determined, normative form of vision. This analysis is considered through three circuits of light processing that have become interwoven to produce distinct discourse networks and realities. The first circuit refers to the waves of photons that penetrate the eye, and the second circuit is triggered by visual memories stored in the brain—these two circuits are the biological circuits that make up visual processing and work to produce the reality that we have become conditioned to. The third circuit belongs to the technological rather than the biological; it is those technologies, eyeglasses and smart glasses, that preprocess visual data to create an illusion of reality before involving the two biological circuits. By preprocessing and manipulating visual information in real time, these technologies create a machinic hallucination that affects our behaviors in significant ways. It is by returning to the precedent of the circuit that a real-time interface between online and offline worlds is most likely to be attained. As in the previous chapters, our genealogical approach gives us the tools necessary to investigate the world-shaping, behavior-shaping, and human-shaping effects of visual media.

From Media Genealogy to Media Teleology: A Note on Our Chapters' Respective Conclusions

When the description of our future becomes our present and our past, we (we critical historians of the present, that is) will know we are on the right track. This would be pure method; description become fact; analysis as presence. What does a predictive media genealogy look like? We don't mean "predictive" in the sense of predictive statistics—a model to map the weather or foresee, within a certain margin of error, the results of a major election. Such naive empiricism is not our metric. Moreover, metrics are not part of our truth-game. Rather, prediction in our proposed frame allows the same imaginative energy required to invent concepts to reach beyond the historical material a genealogy feeds upon.

What will happen when our extensions, the machines of our disembodied desire, grow so powerful that they override, replace, or otherwise ensnare our drives? And do we so-called humans still have access to the switches that govern such circuits? Media teleology, for us, presents a challenge to write from the vantage of the circuit's source of energy. Bernard Stiegler is suggestive on this point in two ways:

> We need to think again of the long term, that is, of teleology. This is because we cannot think of the long term without having in view a fixed/determinate teleological horizon. The problem is how to produce teleology without theologizing.[87]

> Drive, as I said, is not teleological, since it seeks the instant satisfaction of needs, and has no intentionality, in the sense that its aim is immediate, while desire hungers for something that is endless. . . . Industrial capitalism is itself teleological, that is, it has a project concerning society, while finance capitalism is drive-oriented; it is piratical; it consumes everything wherever it operates.[88]

What is the circuit's desire? Or is the circuit animated by a drive? Are the two as distinct as Stiegler would have us believe, or are there two types of circuits, as there seem to be two types of capitalism? We might ask: Once the circuit is enervated with its own self-modulating, self-maintaining flow, how do we treat the circuit's agenda: drive or desire, pirate or project? The perfect machine communicating perfectly with all of the cosmos throughout time; endless. With our chapter conclusions, we make a modest proposal grounded in the imaginary, not to write "fictions," as Foucault (some would say infamously) suggested, but rather to end each chapter with media genealogical speculations. Extending Foucault's position, this does not mean "that truth therefore is absent."[89] Science fiction is often lauded or derided for its relative ability to foretell the future. Our speculations are located between drive and desire, in Stiegler's terms, in that they propose a means for thinking about the world brought forth by drive (a singularly driven logic of immediate concern run rampant) versus one that unfolds according to an unbound telos. Including telos in our repertoire of theoretical tools allows us to add other, imaginary chairs to our list of chairs in circuits. For instance: in another prison, a fifth circuit-laden chair is devised: an electric chair of different sorts, a media-saturated behavior-modification apparatus that was meant to cure criminality. The old prison is simply and permanently antiquated. The difficult process of disciplinary power is short-circuited and simplified through the psycho-chemical persuasion of analog media.

> True enough, one wall was all covered with silver screen, and direct opposite was a wall with square holes in for the projector to project through, and there were stereo speakers stuck all over the mesto. But against the right-hand one of the other walls was a bank of all like little meters, and in the middle of the floor facing the screen was like a dentist's chair with all lengths of wire running from it. . . . And then I found they were strapping my rookers to the chair-arms and my nogas were like stuck to a foot-rest. . . . One veshch I did not like, though, was when they put like clips on the skin of my forehead, so that my top glazz-lids were pulled up and up and up and I could not shut my glazzies no matter how I tried.[90]

This "chair of torture," as Alex would call it, is not a final solution in the humanizing of execution, but rather a means to rewire the circuitry of the criminal brain, consciousness repurposed, cognition changed, though neither carbonized nor discolored. Kemmler's chair and Alex's chair exemplify two different circuits created to solve the same problem: the human propensity to murder each other, or as the Cylons would have it, "man's one true art form."[91] Wired to the prison but slotted differently, Alex and Kemmler are necessary to close their particular circuit, an integral part of the electrical solution.

Kemmler appears as a notable example in a handful of historical accounts, while Alex is prominently ranked the twelfth "Greatest Villain" of all time by the American Film Institute. But where would we rank his machinic cure on that list? Is the "chair of torture" any less menacing a machine than another Kubrick villain, 2001's AI antagonist HAL 9000, who came in just after Alex at number 13? Surely the true antagonism of *A Clockwork Orange* is between human and the all-too-human desire for machine perfection, not some pitting of id versus superego. And these two readings are suggestive of our media teleological analysis. Is the circuit

Alex completes driven by an immediate goal, or is it animated by the transcendent? Is it a simple circuit or, instead, a subroutine in a universal machine?

The crucible through which Alex is reformed, made anew, is a common sci-fi trope.[92] The forging of something transcendent from the materials of the merely human, by the merely human, is a point of cultural fascination riddled with cultural anxiety. The drive for self-perfection is deeply laden in the circuitry of the West, the ghastly outcomes of which appear regularly in sci-fi, and far too often in its history. The ultimate form of such expressions produces a transcendent being simultaneously human and entirely inhuman: completion through annihilation. The Cylons of *Battlestar Galactica* are a prominent exemplar.

The mythos to emerge from this telos is the eternal return of transcendence. After untold thousands of years, the crew of *Galactica* returned to Earth to start their civilization once more. The place of their own birth, and hence the Cylons' as well, promised a beginning and guaranteed their undoing. The two are inseparable, as both originate and terminate at the same point in an endless cycle of rebirth and destruction. The teleology of the circuit is in the end the teleology of the human/Cylon; circuitous encircuiting. Heidegger's assessment of the eternal return is based upon his reading of Nietzsche. He suggests that the real burden of the eternal return is that it is thinkable and as such cannot be unthought. Nietzsche's aphorism from which Heidegger draws is succinctly titled *The Heaviest Burden*. A few lines provide summary:

> If that thought acquired power over you as you are, it would transform you, and perhaps crush you; the question with regard to all and everything: "Do you want this once more, and also for innumerable times?" would lie as the heaviest burden upon your activity!

Humans have no choice but to grapple with the possibility that they too are circuitous, hardwired to repeat over and over again the mistakes of their own undoing. "Do you commit yourself to this circuit-cell which promises eternal singularity?" In the end your answer may not matter, for the circuit chooseth, storeth, and taketh away.

1

How to Make a Soldier into a Medium

Docile Bodies in the Signaling Circuit

The subject is a particularly troublesome kink in the wire in terms of transmission. As a starting point, to articulate the problem from the circuit's point of view: the subject is not a user. The subject cannot be a so-called human. The encircuited subject must be a cybernetic non-unity, a moment of subsumption rather than an entity that persists outside of the circuit's time/space. Friedrich Kittler's approach heavy-handedly insists on an erasure of the subject; it posits the subject as mythos. A genealogical turn, however, leads us to insist upon something more complicated and more insidious. Kittler frames his two historically situated media apparatuses as "discourse network 1800" and "discourse network 1900"; the former is literate and meaning-centered (i.e., semantically shifty), and the latter is digital and logic-centered (i.e., not semantically shifty).[1] On the 1800 end of the continuum, the subject is speaker, writer, reader, author, learner, interpreter. On the 1900 end, the subject is only sender or receiver: user. In both of these models, the media conduit is also literally made of human bodies as media objects in the form of scribes, telegraph and telephone operators, and the like.[2] Sybille Krämer, in response to Kittler's subject erasure, calls the moment of crossing between human and nonhuman media "mediality," and names the subject-medium interface the "messenger."[3] With the circuit, however, we want to talk about a subject that doesn't ever "get" the message, but rather a moment of mediality that gets the message through.

This chapter focuses on the founding of the United States Signal Corps during the U.S. Civil War[4] as a step toward so-called humans' willing co-construction of a circuit made of communication technologies that would eventually render both the human and "communication" effectively obsolete. This moment specifically involved engaging the body in a technical media apparatus in order to appropriate and then subsume it. It is a moment of mediality,[5] a moment in which the messenger tries to disappear by becoming the message. We describe the moment as an especially important one in the circuit's problematic relationship to the subject, and we do so via a genealogy. Media technology didn't just erase the so-called human body in one easy jump; rather, media *incorporated* the human body; media technology required the human in order to constitute itself. Foucauldian "docile bodies," we argue, can be productively theorized as technical media that are the direct result of a problematization responsive to specific relations of power and relationships to knowledge. A host of media have depended upon disciplined human laborers to code, process, and repeat messages that to them look as much like "eye wash" as sequences of 1s and 0s that inhabit RAM or the flashes of light illuminating fiber optic-cables.

The telegraph—and, in the Signal Corps's case, flag telegraphy—is a step in the conversion of communication from a system of "literary," time-bound symbols into a system of discrete, unambiguous numbers. In a sufficiently critical history of the circuit-ensnared present, the Shannon–Weaver model of communication can be conceived of as technology's own aim—almost an ethos of the media-as-agent itself. Moreover, in such a history, the body is ultimately an *imperfect* medium—an intermediate medium, a way station on the teleological track to perfect transmission, to the disappearance of the human. The intermediate dependency on symbolic systems—the human mind as processor, a semiotic system as signal—is imperfect, as well, from a Kittlerian point of view, because Kittler's teleology cannot contain something so temporally dependent and ephemeral as the moment of mediality. The flag-based semaphore system is one body-bound, time-bound step on the road to war-oriented, totalizing, always-"on" film/gramophone/typewriter-type technologies (e.g., radar). Taken this way, moments in the story where bodies are materially tangled in media apparatus become crucial ones for critical study.

One step toward the encircuited subject is the widespread enculturation of a digital logic in order to ultimately render transmissions readable and actionable by nonhuman systems. When we talk about a "logic of the digital," we do not necessarily mean "digital media" per se. We rather mean to evoke a way of being and knowing that has existed across history and that requires and therefore enables what will eventually become digital media as they are commonly understood. Our evocation of a digital logic is in keeping with Kittler's tracing of the semantically slippery, meaning-rich mother tongue (discourse network 1800), through the inscription and protocol-privileging literate media (discourse network 1900), and into the digital age beyond. When we say practices follow a digital logic, we are simply noting practices of mediation which require discrete, modular, nonsemantic inscriptions for transmission.[6] A digital logic is an epistemic requirement of the circuit, the circuit's way of solving the problem of subjective mediality. In its ultimate configuration, circuits will have circumvented the human body, the moment of mediality, and the previously human messenger will be a nonhuman stand-in—a software-enabled algorithm. The body will be, in the end, more material to circulate, the contents of a car, a hospital bed, or a shipping container. In the interim, circuit-driven technology trains and makes use of the body (rendering it a docile body, per Foucault) and the circuit subsumes the subject. To be clear: it doesn't erase the subject. The subject coexists inside the encircuited docile body, but it has, willingly or unwillingly, knowingly or unwittingly, given up its access to the switch.

Theory and Telegraphy: Carey, Peters, Kittler, Hayles, Krämer

The telegraph over the past four decades has become a touchstone for media historians as a way to pinpoint a host of social, technological, economic, cultural, and military changes.[7] Kittler writes, in "The History of Communication Media," that "the electric telegraph, optimized on the basis of letter frequency and charged by the number of words, was the first step on the road to information technology."[8] Furthermore, for James Carey it represents a moment that helps us understand that "models of communication are then not merely representations of communication but representations *for* communication; templates that guide, unavailing or not, concrete processes of human interaction."[9] In simple terms, the advent of the telegraph led to a reorientation of what communications was supposed to accomplish, what problems it imagined to solve. Further, as Carey argued, if scholars were only interested in producing research that measured the apparent instrumentalist effects produced by discrete messages, then our understanding of what communication can and should accomplish would continue to be skewed toward the desires of commercial enterprises and geopolitical actors to exert control over greater territory. From Carey's vantage, Kittler's focus on a transmission model, based in a military telos, might be said to lead to a military ethos. According to Kittler, media overcome time by precisely recording and preserving otherwise fleeting events. According to Carey, continuity in time is also accomplished through repetition of actions and phrasings, not merely the technological archiving of all that happened. In that sense, Carey's ritual model, when combined with our focus on the disciplined body of telegraphy, gets us to a different approach to time-axis manipulation. And time-axis manipulation, according to Krämer, is the key problematic that Kittler provides media studies—his "main point." Indeed, the explanation of the technological as a modality of time management is precisely the "main point." The most basic experience in human existence—and this is relevant because the human is, after all, a physical being—is the irreversibility of the flow of time. Technology provides a means of channeling this irreversibility. In media technology, time itself becomes one of several variables that can be manipulated.[10]

For Carey, time-axis manipulation is also accomplished in the body, not via technology per se but via a technology of the self. Time-axis manipulation is not imagined as a media storage container but rather as a repeated bodily process through which culture is maintained by intense disciplinary techniques passed down over generations. Carey uses the example of highly disciplined boys' choirs whose training has been consistent throughout the Middle Ages and beyond.[11] Communication's power is to maintain meaning over time, not merely discover the atemporal in communication, as with Kittler's discovery of digitality in Greek music. It is not the maintenance of ideal forms (numbers) that exist in music; rather, cultural ideals are trained into the body, inscribed in the neural networks through repetitive training. The body is not merely collector and sender of signals, but becomes the storage mechanism of culture. While Carey was interested in how the Catholic Church used discipline, we turn instead to military culture and the ability of the soldier's body to speed up transmission over greater distances.

One major Kittlerian move, the one that intellectual biographer Geoffrey Winthrop-Young calls Kittler's "war answer," maintains that "the media a priori [is] collapsed into a martial a priori."[12] Kittler's "war answer," transmission model orientation, and digital teleology come together in one phrase that we will work to enrich: "Command in war has to be digital precisely because war itself is noisy."[13] Circa 1860, the electric telegraph is the arbiter of digital military signaling at its best, for it creates a mechanism for binarily coded messages—semiotics reduced to dashes and dots. At least in the conceptual dimension, noise is filtered out. Kittler uses the German invasion of France in 1870 as a decisive moment in which the telegraph comes to reorient the distinct division of military signals from troops. Military commands are finally freed from the necessity of humans (or trained animals) to physically move texts through space. Troops are moved by train, while signals are moved along the same vectors by telegraph lines that run alongside the tracks. Kittler is by no means the first to emphasize the telegraph as the technology that separates communication from transportation. Carey's well-known article "Technology and Ideology: The Case of the Telegraph" marked out this territory as fertile ground for understanding a host of fundamental alterations to both how communications was conducted and how communications is conceptualized.

In a 2006 essay, John Durham Peters revisits Carey's analysis by way of Kittler, suggesting that by adding Kittler's concerns with time/space manipulation we get a very different story in which the "graph" (inscription) takes precedence over the "tele" (transmission). Peters suggests that by invoking Kittler and other German media scholarship that focuses upon the machine rather than the human, the central role of telegraphy in "giving birth to new kinds of scientific instruments, in subdividing time, in anticipating the processing (switching) infrastructure of the modern computer, and in inspiring Einstein's special theory of relativity"[14] will be understood. We would like to expand the scope of historical focus on this point in the history of technology by looking directly at the years just prior to Kittler's 1870, 1860 to 1865 to be exact, and "thinking with" Kittler in order to differently animate how we might conceive of telegraphy, digitality, and what we are calling the "military science of the sign." Specifically, we would argue that the "war answer" should be extended into "the war answers." Debates within the military during the Civil War over the effectiveness of flag telegraphy versus mechanical telegraphy are telling in exposing how favoring spatial over temporal dimensions of military strategy leads to different discourse networks. Where Carey links trains, telegraph wires (materials), Morse code and semaphore (system of signs), and the "communication" of market materials and human bodies (culture) by a timetable (bias) as an ideology-defining moment, we return to the same context/moment in time to trace the linkages of different materials (bodies) to similar semiotic systems (telegraphy) for different ends: war.

The adoption and practices of electronic telegraphy has long been seen as a seemingly irruptive moment in the history of communication to be revisited and enriched. David Hochfelder's comprehensive history, *The Telegraph in America: 1832–1920*, begins with the Civil War. Hochfelder's account, however, focuses almost entirely on the civilian United States Military Telegraph Corps, its tactical benefits to the Union effort, and its effective consolidation of telegraphy infrastructure, personnel, and practices which translated into civilian industry and culture when the war ended. N. Katherine Hayles, in *How We Think: Digital Media and Contemporary Technogenesis,* uses commercial telegraph codebooks from the mid-twentieth century as a case study with which to illustrate how "technogenesis implies continuous reciprocal causality between human bodies and technics."[15] "Before the balance tipped towards machine cognition," Hayles writes, "a zone of indeterminacy was created between bodies and messages."[16] Hayles explores that zone in terms of how Morse code was used to send the telegram, and relates it to the technological imaginary (or "technological unconscious") that enabled the cryptography that drove Alan Turing and the allies at Bletchley Park during World War II. We see flag telegraphy in the United States a century earlier as very much part of the same story. It is key to the work of historians of technology and media, however, to extend our view into that "zone of indeterminacy" such that the balance-tipping is less of a break and more of a continuous line.

Key to our treatment of flag telegraphy's moment of discourse network (inter)mediation is Kittler's distinction between "technical media" (what Hayles calls "machine cognition") and the accepted understanding of media in general. Media are, of course, a means of delivery, storage, and logistics for the purpose of communication. Kittler's technical media, though, are digitized messages for the contemporary computer—the kind of media that human experience can't "read" at the level of signal. Technical media, according to Kittler, "make use of physical processes which are faster than human perception and are only at all susceptible of formulation in the code of modern mathematics." The messages of technical media, says Kittler, are information "decoupled" from communication. But Kittler also notes that technical media predate computing. "There must always have been technical media," he writes, "because any sending of signals using acoustic or visual means is in itself technical. However in preindustrial times channels such as smoke signals or fire telegraphy which exploited the speed of light, or bush telegraphs and calling chains making use of the speed of sound were only subsystems of an everyday language." We would argue that such proto-technical media always already operated by a digital logic. Civil War flag telegraphy resides there, in Hayles's "zone of indeterminacy," in the mess of the proto-technical media driving the digital logic—a noisy place made of smoke, bodies, feathers, canvas, and wire, all carrying a signal that longs for the clean fiber-optic cable.

A Military Science of the Sign: The Invention of the Signal Corps

A northwardly facing uniformed man stands on a bombed-out hill rapidly waving two distinctly marked flags, each affixed to a twelve-foot pole. Mostly hidden behind an embankment and dodging enemy fire, well beyond the range of unaided sight, a second man intently stares at the first through his field glasses—a mechanism consisting of two identical telescopes that provide stereoscopic vision at great distances. Without looking down, his hands frantically work a quill pen marking out a sequence of seemingly indistinct numbers and letters. Tearing his eyes from his glasses as he turns around, he leans down, picks up his own poles, and wags them to and fro in a systematically snapping display. Equally far north, a third man processes this same sequence of signals. From the battle lines of the Deep South, this message is repeated and remediated until it reaches the War Department in Washington, D.C., as a telegram, where Abraham Lincoln stoically awaits, nodding acknowledgment to the telegraph clerk who delivers it—now a scribbled note on War Department stationery.[17] But the Great Orator is not concerned with rhetorical flourish today. Command in war, he knows, should be noiseless, unambiguous, and precise.

Dr. Albert James Myer, a Union officer in the Civil War, was the first signal officer and (as history has since named him) founder of the United States Signal Corps. He developed the military entity in parallel with the establishment of the U.S. Military Telegraphy Corps (USMTC), a civilian organization devoted to wire telegraphy, and the two organizations did not communicate much at the outset. In fact, Hochfelder notes that the disconnect between wire telegraphy and the other forms of military signaling was a perennial problem for the Union.[18] It is not merely by chance that the military needed Myer or that Myer needed the military. In 1844, at the age of sixteen, Myer learned how to work a telegraph while at Hobart College. Four years later he completed a degree in medicine and made a special study of sign language for deaf mutes. At the time of his studies, Myer worked as a telegrapher at the New York Telegraph Company. His thesis developed a binary sign language that gave bodily life to Morse code. As his thesis attested, "It is not strange, therefore, that under such circumstances, I should not conceive the idea of aiding, with so simple a speech as can be founded on this principle, those whom the Deity has seen fit to deprive of the natural organ."[19]

Myer's story could lead to a Kittlerian reading; Kittler maintains that "handicaps isolate and thematize sensory data streams,"[20] and Kittler's story of the phonograph leads us back to Charles Cros, who worked at a school for the deaf, to Thomas Edison, whose hearing impairment aided in the first captured vocal data as he screamed into the bell mouth of his phonograph, and to Alexander Graham Bell, whose mother's and wife's deafness led him to focus his scientific efforts on sound technologies and ultimately invent the telephone. Kittler further points to the first typewriters as a technology built by the blind for the blind.[21] Further, the story of Myer as founder of military communication systems ties Kittler's "war answer" to a Kittlerian explanation of media as responses to the human body's inadequacies contra the circuit. Yet, as Mara Mills and Jonathan Sterne point out, "Kittler may be correct that early technical media were developed 'by and for' deaf and blind users . . . he ultimately reduces the significance of this point to passive illustration." Mills and Sterne argue that such ableist epistemologies fail to acknowledge the norming and normalizing understandings which Kittler's and Myer's "handicapped" or "deprived" bodies are built upon. Instead, they suggest a media history that acknowledges "norming in science, technology, and medicine."[22] We would like to add the military to such a list.

Kittler famously claimed nothing is that which cannot be mediated, and Myer's theory of mediation was at least as far-reaching as Kittler's. "There is no thing or sight or sound or motion or taste or odor perception sensation or indication," Myer writes in his doctoral thesis, "but by which or through which ideas and meanings may be intelligibly transmitted and which may thus be used for signal communication."[23] His system for the deaf binarized signing into two simple, distinct movements: one equated with a dot, another a dash. Myer suggested that sight, hearing, and touch could all interchangeably be made to sign with equal semiotic efficiency. As Kittler would clarify 130 years later, once everything is made binary, "any medium can be translated into any other."[24] For Myer, the corpus of human sensation was an untapped cornucopia of semiotic capacity.

Myer conceived of his signal system for the military—an application of the same concepts that he had used in his dissertation—while he was serving at Fort Davis in Texas.[25] According to the U.S. Army's official history, Myer saw Native American scouts using "crude" flags to signal and realized that a visual signal system made sense for communication over long distances in the wide-open terrain. He apparently developed, deployed, and promoted his system himself—he taught it to his own unit in Texas, and it was authorized by the War Department in 1858. Myer was made signal officer—a position invented for him—in 1860, and as soon as the war broke out he positioned himself and the Signal Office as an important Union advantage over the Confederacy. As soon as shots were fired at Fort Sumter, Myer asked for a command of ten officers, and specifically "three intelligent and able-bodied men who [could] read and write."[26] His ideal corps trainees would be hyperliterate semiotic experts, able to turn any sense—a body's perception of the material—into a system of signs. The signal officer's body must become a signal producer (hardware) and data processor (software) for an overground signaling machine.

Making Docile Bodies Media: Signal Corps Training and the So-Called Human

In *Discipline and Punish*, Foucault uses the disciplined soldier as the ultimate example of what he comes to call a "docile body": "the human body . . . entering a machinery of power . . . that [the machine operator] may operate as one wishes, with the techniques, speed and efficiency that one determines."[27] He applies these concepts to the precise choreography of a soldier's motions in a weapons drill. The book *What You Should Know about the Signal Corps* (a popular publication from Norton in the 1940s) calls the process "the conversion of new soldiers into competent craftsmen of war."[28] The Signal Corps manual of 1898 notes that "the signalist, once well taught, becomes thereafter independent of signal books, or codes, or especial apparatus; and, in a life of active service, may never encounter instances in which he can fail to open communication with one similarly taught, if both are signal distance, giving attention, and provided with the means for operation."[29] The related metaphor of soldier *as* weapon is now familiar to the public imagination; the Navy Seals, mythic "trained killing machines," are programmed assemblages of guns, boots, arms that hold the guns, feet that fill the boots, and minds that are steeled to do what must be done. The soldier's docile body and weaponry assemblage is, in effect, a weapon.

Our argument is similar, but with a focus on the concerns of communication and transmission; the Signal Corps soldier as docile body is a type of proto-technical medium by minimizing noise at the medial moment.

Figure 1.1. Basic wigwag, signaling numbers. Joseph Willard Brown, The Signal Corps, U.S.A., in the War of the Rebellion *(Boston: U.S. Veteran Signal Corps Association, 1838), 97.*

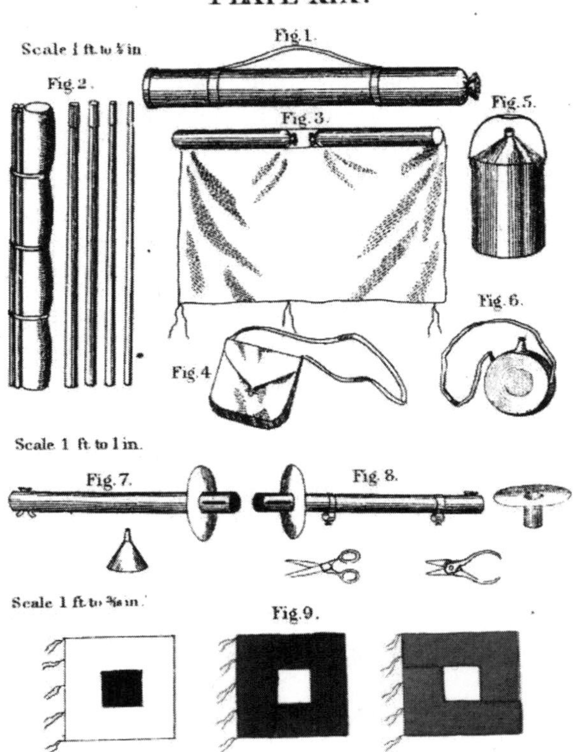

Figure 1.2. The Signal Corps officer's kit, circa 1905. Albert James Myer, A Manual of Signals *(New York: D. Van Nostrand, 1866), 176.*

The soldier's body becomes media by a disciplining (discipline in the Foucauldian sense, programming in a Kittlerian frame, and/or encircuiting in the larger story of this volume) at the level of material, semiotic, and cultural processes. Hochfelder notes that the civilian officers of the USMTC lacked such programming and "often refused to conform to military standards of discipline."[30] Later, after the Civil War, after Myer's removal from leadership, just before the turn of the century, the U.S. Army issued an instruction booklet detailing the property and general regulations of the Signal Corps, which institutionalized the soldier assemblage by detailing the materials used and the actions performed in the communication machine—ostensibly training the soldier how to be a moving piece inside of it. (We should note that even though it is part of a machine, the docile body is not a weakened figure. Disciplinarity is productive and doesn't merely tame the body. Rather, the body is given new communicative capacities; its powers are amended.)

The thusly encircuited soldier–medium's materials are his body—particularly his arms—and his standard equipment. Myer's manual breaks the signalman's equipment into three groups: the kit case, canteen, and haversack. The Signal Corps soldier's standard kit in Myer's day contained five basic canvas flags for flag telegraphy, a set of two flag poles (adjustable like tent poles), and a torch. The canteen contained torch oil, a funnel for filling the torch, and scissors for trimming lantern wicks. The haversack contained a writing kit made up of a notebook and pencils (for inscription and storage of information/messages), extra wicks for storage and a sewing kit (for maintaining signaling technology), a screwdriver, pliers, a telescope (for extending the vision/overcoming space), code wheels (for encoding and decoding, i.e., processing capacity), clean rags, and a compass (a location-finding technology). Later, in 1898, the Army published a booklet of *Property and General Regulations of the Signal Corps* that was thirty-three pages and seventy-three entries long, detailing the care, storage, and transmission of the corps's material property from soldier to soldier.

But those are just the nonhuman materials. The most important material medium—the transmitting, receiving apparatus—of flag telegraphy is a set of arms and legs to wigwag, a pair of eyes to see, and a mind to decode; the system's semiotic choreography is absolutely dependent upon the soldier's body. The body's arms move the flag and are the primary (though impermanent) inscription. The body's coronal plane (its line of symmetry) becomes key to the system's binary code. The body's eyes become the receiver, the body's brain becomes the processor, and the body's hand records the interpreted message for storage (on the material, nonhuman medium of paper). So the entire process of remediation—from what Myer himself calls the "transient signal" to a "permanent signal"—depends upon the soldier's body and the soldier's enculturation into an apparatus with digital logic.

The Signal Corps's methods of mediation are also dependent upon the soldier's aptitude for language; signal science is semiotics, and Myer's signs deployed a new, logic-oriented literacy. When Myer enlisted his first group, he required that all recruits be able to read and write (a tall order to fill in the 1860s). They read and wrote in the mother tongue, though—the natural language of discourse network 1800. In order to function in the technical media proto-circuit, soldiers had to "speak" in several different species of numerical code, the most ubiquitous and useful being a "general service code" described in Myer's "Field Signals by Two Elements," which is the basis of the corps's wigwag choreography. General service code is binary; it uses the two sides of the body as the dividing line between "this" (one)—a wave of the flag or torch to the left, and "that" (two)—a wave to the right; the code uses recombinations of this and that to build more complex messages. Myer's "wigwag" system for his first unit included a dip to the front that stood for three. In the initial system—before messages became encoded to avoid enemy comprehension—sets of numbers stood for letters: A was 11, B was 1221, C was 212, and so on.[31]

The first half of Myer's manual is a primer to this end, and here the Kittlerian media scholar will note that Myer's system is still semiotic—entirely dependent upon the sign. But the same scholar should also note that Myer's system is *designed to avoid semantic slippage*, dedicated to a digital logic, through detailed discipline. Myer recognizes and strives to minimize the problem of processing "noise" due to human inscription, sending, receiving, and processing. On the one hand, the Signal Corps and its first set of technological apparatus for communication were examples of literate media—part of discourse network 1800. The codebook, code wheel, and writing kit limited Signal Corps enlistment to a certain literate class, for example, and transmission was entirely dependent upon embodied experience. At the same time, the system doesn't rely on natural language; it is binary—a sign of what lies well beyond the encroaching discourse network 1900. Myer's general service code even requires a proto-markup language in order for the human processor to make sense of messages—a set of signals at the beginning of the message that told the decoder how to decode the message, and in what order, and a set of signals to sign off and/or change code systems, and a sign-off feedback dialogue that rechecks the accuracy of the message. Myer's design problem is the same, then, as Claude Shannon's, with human bodies as signaling technology and processing power, and with a protological military culture as its software.

Possessing the materials required to send and receive signals and becoming a signalist fluent in one or another system of signs doesn't make a body a media technology, per se. The last layer is a very specialized enculturation—a discipline. The process by which a man is made into a soldier—the disciplining of soldiers—relies on what Carey calls ritualistic communication. From a Kittlerian point of view, ritualistic enculturation is a sort of programming, a preparing of human, cultural/behavioral software on which the human and nonhuman hardware (body/flag assemblage) runs. The enculturation of a soldier is a much-documented process. The 1898 *Property and General Regulations of the Signal Corps* itself even details the behaviors related to its own changing-of-hands: "These regulations are only issued for official purposes, and will be transferred by signal officers and others receiving the same to their successors when relieved from their property accountability."[32] *A History of the U.S. Signal Corps* quotes Captain Henry S. Tafft reminiscing about his education as if it were an induction into a secret society: "I was at once initiated . . . into the weird mysteries of aerial signals with wands or motions of any kind, with flags by day and torches by night."[33]

(Proto-)technical Media: Smoke, Flags, and Composition Pistols

In her exhaustive history of the Signal Corps through the 1990s, *Getting the Message Through*, Rebecca Robbins Raines describes the conditions at the corps's inception in the following terms:

> The [dispersed situation of the U.S. Civil War] demanded new methods of tactical signaling, and Myer's wigwag system helped to bridge this communication gap. Before the invention of the telephone and the radio, the soldier possessed no tactical communication device that he could carry onto the battlefield. Wigwag, despite its limitations, enabled signalmen to communicate between prominent points on or near the battlefield and the commander's

headquarters. The portability of the equipment permitted its use on horseback and on shipboard as well.[34]

The Signal Corps had to patch together the missing links between the troops on the ground (or in the sea) and the existing civilian infrastructure maintained by the USMTC. The media devised to solve the Signal Corps's unique set of problems were designed to overcome time and distance (thus visual flags or sonic telegraphs), issues of mobility (thus modular and attached to the railroad), secrecy (thus encoded), and precision (thus binary/mathematical). All four of these exigencies make the Signal Corps story a site for Kittler's style of historical inquiry. The speed of communication is increased by "exploit[ing] the speed of light . . . or the speed of sound."[35]

Electric telegraph lines follow railway lines, and Signal Corps units are ultimately modularized quite literally as train boxcars. Secrecy requires the encryption of messages (to be discussed later). And the problem of precision requires the "either/or exercises" of screening technologies:[36] important/unimportant, friend from foe, one signal from the other, parsing and triangulating space, breaking down information into the communicable. These all privilege the Shannon–Weaver model of communication and operate by a digital logic by quite literally rendering com-

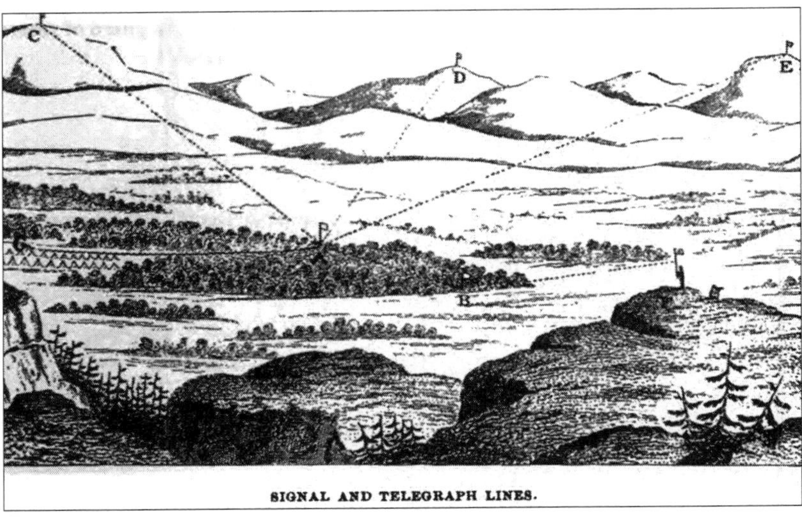

Figure 1.3. Signal and telegraph lines in a landscape. Joseph Willard Brown, The Signal Corps, U.S.A., in the War of the Rebellion *(Boston: U.S. Veteran Signal Corps Association, 1838), 129.*

plex and often ambiguous sociocultural states into simple, unambiguous, binary ones. Messages weren't meant to comprise subtleties of interpretation; rather, they must convey a single command with a singular meaning. Kittler calls this the decoupling of information from communication.

Myer's emergent Signal Corps needs to transmit information with a minimum of material and interpretive noise, but its historical milieu just barely predates the practical application of film/gramophone types of technologies that capture and preserve information without a human intermediary. Later, radar and other screening technologies will be enabled by modern computing, but at this point, in an uncomfortable place between discourse networks 1800 and 1900 and on the way to contemporary technical media, the story is enlivened by a panoply of media, media technology, and remediation/media practices for media historians to begin to consider, especially in terms of seeing mediation outside of its typical suite of related communication technologies and as assemblages of those technologies. Here, flame, bodies, birds, and mountains become parts of the media apparatus.

Visual simultaneity was one answer to the time/space question that didn't depend on electronic sound in wires; it used the speed of light as its transmission. The sight line as "wire" called for both extension of the visual and extension of the vision. The first proto-technical medium that extended the visual by, in Kittler's terms, "exploit[ing] the speed of light" was smoke. As mentioned, Myer's experience with Native American smoke and hand signals in the western territories gave him the idea for communication at a distance with visible/visual media extensions. The cutting-edge communication technology of the time was, of course, the telegraph, but running wires from control centers to strategic parts of the always-moving, never stable or safe military front was always a problem. Myer's flag communication, then, was a kind of line-of-sight, "wireless" telegraphy—in fact it was called "flag telegraphy." It conceptually combined two media—smoke signals (visual connection) and the telegraph (sonic connection)—and their compatible system of signs (a Signal Corps–specific code) to avoid the material noise that the smoke entailed and to overcome the space-related obstacles that laying wire for telegraphy entailed.

The exploitation of the speed of light to carry a signal with the human eye as receiver naturally entails fog, terrain, and the dark of night as systemic "noise." Both the visible and the body's vision had to be extended to reduce noise. Telescopes and field glasses were both a part of the signalman's basic gear. Robbins Raines notes that the rolling motion of a ship rendered telescope technology impractical, making field glasses an invaluable vision extension at sea.[37] The design of the wigwag flags themselves addressed the problem of visual noise by being bright white and red and inversely patterned (with a red center square in white on one flag and a white center square in red on the other) to be easily distinguishable from one another. The signal torches were the next solution, for use at night. Signal torches included a long signaling torch, which had a shade to prevent the flame from being disturbed by wind or sweeping wigwag maneuvers, and a ground torch, which served as a point of reference in space for the observer. Signal Corps historian Joseph Willard Brown (1896) describes even more complicated ways of overcoming distance and terrain-related noise. "Chromosemic" rockets "attain great elevation, and are sometimes seen when signals made on the ground would be unseen."[38] Brown also describes a "signal pistol" fitted with "composition fire" cartridges that emit colored smoke in the daytime and colored flame at night.[39] Kittler describes the Remington typewriter as a "discursive machine gun,"[40] yet here is a firearm that very literally composes.

Architecture and terrain are, in this system of mediation, at once sources of noise and means of overcoming it. Mountains, steeples, houses, rooftops, scaffolds—the landscape itself becomes infrastructure for the Signal Corps apparatus, the discourse network. Robbins Raines notes that "careful site selection proved especially important to avoid such obstacles as dust or rows of tents."[41] She also quotes Myer's manual, which instructs the signalman in how to use trees as ad hoc signal stations:

> The flag-man may then secure himself in the tree with a belt or rope. The officer fixes his own position at some other place in the same tree, and rests his telescope among its branches; or what is better, ascends another tree for this purpose: as the first is apt to be so shaken by the motions of the flagman, as to disturb the vision through the telescope.[42]

Robbins Raines goes on to describe how specially constructed towers, tall buildings, and church steeples all made good bases for signaling and signal receiving. A hill near Gettysburg, Pennsylvania, called Round Top became the most famous signal station of the war. One Confederate general called the hill "'that wretched little signal station' and always tried to keep his artillery 'out of sight of the signal station on Round Top.'"[43]

The Signal Corps saw its first action in June 1861 at Fort Monroe in Virginia, and the wigwag media technology met the material realities of the landscape it was designed to overcome. Signal officers used bridges at Fort Wool for signal posts, as well as boats to scout the locations of Confederate guns in Fort Monroe. According to Washington's *National Intelligencer*, "General Butler . . . gave the directions by signal for elevating and lowering the range of the gun so as to strike particular points. The result of which was satisfactory in every instance."[44] Thus signaling overcame the "natural barrier" of the sea and was used to monitor the launching of new projectiles from cannons (a kind of guided missile). Here we see Paul Virilio's thesis, followed up by Kittler, coming to fruition. Troops need visual knowledge to act from afar. They need to understand the terrain and placement of the enemy beyond their immediate perception. Combatants call upon "optical media," in Kittler's terms, to formulate their "logistics of perception" for "technical warfare," to use two of Virilio's phrases.

Mapping time/space coordinates for strategic advantage is the ultimate goal of military knowledge. Logistical perception is more than simply a prosthetic addendum (like the telescope); it is also a more general technological apparatus that coordinates military planning, engineering, architectural structures, means of communication, and, at least until weapons automation began to appear in the 1950s to secure the future of the subject-free circuit, military-defined "able-bodied" so-called humans. We can see in the very first battles of the Civil War, during which the Signal Corps is center stage, that the corps is working to create proto-guided missiles. They are providing real-time assessment, coordination, and guidance for newly minted cannons that can fire well beyond the limits of human sight. Cannon fire must be guided from afar, via communications-oriented technical media, to have any tactical or measurable effect.

Secrecy and Precision: The Need for Disembodiment

Within days after the first battle, Myer began work on sending telegraph cables into the sky via hot-air balloons to extend the scope and hence practical range of cannons. Gravity and flora couldn't be overcome. Telegraphy failed to become ethereal until Marconi later made it spectral (and profitable).[45] However, the corps did create aerial photography and, more generally, became the official photographic arm of the Union's military. Virilio acknowledges that the hot-air balloon/camera couplet was a precursor to more far-reaching applications that would be achieved in World War I with the airplane/camera couplet. We suggest that real-time communication is of equal importance to the time-axis manipulation provided by photography.

Hot-air balloons did not feature electronic telegraphy, but they did host wigwag. As Myer readily argued throughout his time with the Army, the lightness of communication technology was often more important than the speed of transmission—hence wigwag, in some contexts, had benefits related to mobility that made it more highly valued than the wire telegraph in the latter's early days. Kittler makes much of the telecommand utilized in the 1939 blitzkrieg of Europe and describes similar problems. Surely nothing like remote command by telegraph, or "telecommand," had been witnessed before. Yet, telecommand optimally demands "tele" command, and spatial knowledge. As Kittler unsurprisingly notes, following Martin van Creveld's (1985) *Command in War*, the most prominent figures in working out how to orchestrate blitzkrieg via radio-enabled telecommand were two German soldiers who had cut their teeth as signal officers in World War I. Norman Ohler and Shaun Whiteside suggest that bodily manipulation was still essential even to the mechanization of blitzkrieg.[46] They argue that the compulsory use of methamphetamines by German soldiers produced the conditions under which blitzkrieg was first spontaneously enacted.

However, as with any open communication system—radio or wigwag, for instance—the only way to make it work for military purposes is to close off the system to enemy eyes and ears. In Myer's milieu, Union officers mistrusted the signal system:

> Signal security posed a serious problem. The chief of staff of the Army of the Potomac, Maj. Gen. Daniel Butterfield, expressed this concern during the battle of Chancellorsville when he ordered that signals not be used because the enemy could read them. Capt. Benjamin F. Fisher, chief signal officer of the Army of the Potomac, lamented in his report of the battle that "the corps is distrusted, and considered unsafe as a means of transmitting important messages. It is well known that the enemy can read our signals when the regular code is used."[47]

So for the same reasons that the Germans developed Enigma several decades later, a code eventually broken by the Allies, Myer patented his cipher disk, which became part of all Signal Corps members' kits.

According to Robbins Raines, the Confederate army never broke the Union code. Still, General Sherman did not place a great deal of trust in the Signal Corps. In his memoirs he commented that he had "little faith in the signal service by flags and torches, though we always used them; because, almost invariably when they were most needed, the view was cut off by intervening trees, or by mists and fogs." The notable exception was at Allatoona "when the signal flags carried a message of vital importance over the heads of Hood's army." Sherman placed his faith in the magnetic telegraph and felt that the commercial lines would "always supply for war enough skillful operators."[48]

One major material issue with incorporating the telegraph into the infrastructure of the Signal Corps was the time and training required to employ the cipher disk to encode, send, and decode telegraphic messages. Telegraphic messages were painstakingly tapped out by a human hand, letter by letter, in Morse code–like pulses with one switch; they were then received, recorded, and decoded by a soldier at the other end. In Kittler's terms, the human-interpretive discourse network 1800 literacy was an obstacle to the noiseless, time- and space-saving transport model of communication. In Krämer's terms, this iteration of the Signal Corps had the goal of a cybernetic system but was overdependent upon layered medialities. Myer ultimately sought to ameliorate the problem by inventing a technical medium. He incorporated his cipher disk into a telegraph machine with a magneto-electric generator. The Beardslee Telegraph (named for inventor and Signal Corps officer George Beardslee's magneto generator), in theory, would allow for a message sender to simply dial in the letters he wanted, which the machine then converted into encoded pulses, and the receiving machine then output in cipher-dial letters.

Myer's vision was to equip all Union army trains with a Beardslee device. As is often the case, the fact of the matter turned out different from the theory. The hand-crank-generated power of the Beardslee was only effective at transmitting a signal some ten miles, and the translation of disk to code within the machine was tricky in the early prototypes, making message sending even more complicated for the soldiers at either end. Although the Beardslee Telegraph was used successfully in the Battle of Fredericksburg, Lincoln's secretary of war, Edwin Stanton, ultimately called

Myer's vision an "expensive failure."[49] Myer's obsession with streamlining the materials in the "machine" that was the Army's signal-sending apparatus was ultimately his downfall. Stanton relieved him of his post in 1863, tired of Myer's continual insistence that all communications infrastructure, including the civilian USMTC personnel, be placed under his purview.[50] From the media historian's point of view, Myer's institutional failure was that his technological vision was before its time. Ultimately, as Kittler claims, "the language of the upper echelons of [military] leadership is always digital."[51]

Although Myer's technical medium—a semiotic machine that converted letters into code and back again—was a failure in the hothouse context of war, Myer's problem was ultimately the same as that of the mathematicians and scientists who worked with John von Neumann in World War II—namely, to provide secure means of encoding, transmitting, and decoding messages as quickly and with as little confusion as possible.[52] The Civil War and World War I are often seen as being a bloody and problematic transition into the era of modern warfare; moreover, contemporary computing technologies are seen in part as instruments producing and produced by modern warfare—that is, computing technologies are one demarcation of what is modern. The "war answer" to the history of computing should certainly include the proto-computing technology of Myer's telegraph machine and the so-called humans who constituted it. Myer, in striving for a clean transmission of encoded information, was driven by the digital logic; he lacked only the technology that another half-century of digital telos would produce.

The Signal Corps Reconciles the Digital

Of course, the history of modern warfare proper, the history that Kittler features in his war-centered history of the digital, depends upon communications technologies that Albert Myer could only have dreamed of. By World War II, advertisements and articles in *Popular Science* and *Life* magazines were recruiting boys to "fight the battle for communication," telling them that "the nerve center of the army needs your skilled hands today!"[53] One of the main attractions of such propaganda is the cutting-edge technology the Signal Corps worked with. A popular nonfiction book from that time, *What You Need to Know about the Signal Corps*, tells a story about the coalescence of technology that isn't far from Kittler's: "It is time to see how the Signal Corps has applied the telescope, telegraph, telephone, teletypewriter, and radio to the communications of the U.S. Army."[54] And the little book's take on technical media and time/space biases? "Modern war," it says, "is a war of vast distances. Signals must bridge those distances. This is a war of rapid movement. Signals must outspeed the fastest plane. . . . Thanks to our modern means of signal communication—the telephone, the teletypewriter, and especially the radio—we have the power to send messages instantaneously to any part of the globe."[55]

The intention of Myer's Beardslee device was realized in von Neumann's MANIAC, whose historians have already traced speed, secrecy, and precision through the combination of big science and the military machine.[56] The institutional entity of the Signal Corps is enmeshed in that story, and the corps's material media—including so-called human bodies, the bodies of pigeons, cloth flags, and puffs of signifying smoke—are, when incorporated into the media apparatus of the corps—necessarily media that operate by a digital logic. The requirements of war are such that the domain of the digital must be built to become more and more separate from the domain of so-called human experience—"a domain," as theorist Aden Evens describes it, "of discrete code, reducing everything digital to its rigid logical mechanism."[57]

The logic of the circuit converts Myer's "dream of the perfect language"[58] into a dream of an existence in which natural language is unnecessary. The circuit's model for communication eschews communication. In the interest of the clean signal, the sign itself is brought under scientific investigation for the purposes of military strategy in what we are calling the *military science of the sign.* The specific goals of military signaling, here, are brought under scrutiny by a physician who had cut his teeth on the problems of signal transfer and translatability. The technological history of the Signal Corps, then, dovetails with the history of the life sciences as well as the history of remote sensing and all of its new territories of scientific knowledge production—photogrammetry, satellite imaging, and light detection and ranging (LIDAR) data sources. Remote-sensing technologies and their respective epistemic territories, the body and the planet, beg consideration in terms of histories of surveillance, governmentality, and biopower.

The Signal Corps also raises several potential theoretical considerations for further productive thought from the standpoint of the history of communications. One is the question of how time/space axis manipulation is about the ability to control space through speed—this is Carey's suggestion—or to control time through the spacialization of data, which is Kittler's and Krämer's. In the example presented here, we could move in two different directions. On the one hand, Peters wants to follow the Krämer line of thinking to broaden our understanding of how control over the temporal allows for greater scientific understanding and for more nuanced control technologies that act upon smaller objects and at decreasing intervals.

We instead follow Carey's line of thinking, but via Kittler's war thesis and the military science of the sign. Peters's line of thought remains, though, and promises to be productive. How did the added speed and subsequent time control of a networked military produce new knowledge and behaviors at the micro level? What changed in the practice of the single sergeant in charge of a division when he was given more and more diverse knowledge about the rest of the military body over space? How did time-bias control over space change management of provisions? What about the soldier's mediality? Were soldiers dehumanized, perhaps, because of the increasing disappearance of time and space and an increased sense of the army as a unified body in which individuals were movable, disposable parts?

A final move is to reinvigorate the idea of human bodies as elements in communicative machines—a maintained mediality in the circuit—via military disciplinary mechanisms. The ultimate example of such a machine can be found in Cixin Liu's *The Three-Body Problem,* in which thirty million soldiers wigwag black and white flags representing 0s and 1s in a "complicated and detailed circuit" that forms "the most complex machine in the history of the world." Though a work of fiction, Liu's analysis of a computer composed of disciplined human bodies was spot-on: "Even though the whole is complex, what each soldier must do is very simple. Compared to the training they went through to learn how to break the Macedonian Phalanx, this is nothing."[59] While all modes of systematic human communication, language and writing, demand a certain element of discipline, training, and precision to ensure accurate data transfer, coding and decoding are only two sources of noise. There are also numerous considerations of imprecision in the medial moment in either the design of letters (Is that an i or an l?), the necessity of translatable clarification (two, to, or too), or the redundancy of synonyms. War asks humans to reduce such uncertainty in a realm where uncertainty rules—hence discipline, which itself is more broadly an attempt to turn the undifferentiated mass into an orderly system. It works to separate according to a binary logic of the normal and abnormal, and its very system of "selective service" amplifies and institutionalizes the norming of "able-bodied," as we will show in the next chapter. Root out the abnormal and temporally ephemeral (noise) to create the perfectly tuned system. But the drive for the circuit's certainty, for a world that speaks to itself, free from the imperfections of human perception and communication, is precisely the teleological nightmare that Carey saw at root in the transmission model of communication. In a very different context, Carey saw in disciplined bodies the ability to transmit not the precise signal of the general but the imprecise meaning of being human. Conceiving of the human in terms of transmission, even in such disciplinary settings, misses another central temporal consideration of media: the maintenance of society by the continual connection of points and circulation of flows in real time. The society that the circuit maintains replaces the value/outcome of human meaning (and, perhaps, the lived experience of mediality) with the value/outcome of signal security.

Media Teleology: Weaponizing the Suicidal Psychotic Break

The military circuit maximizes its productive power and minimizes its security risk by keeping the soldier, especially the soldier's body, as a (albeit medial) medium, never a messenger or message. We do not, like bad readers of Kittler, pretend for the sake of theoretical continuity that the body must be kept separate from the subject, however. On the contrary: in an adequately political media genealogy, the subject and the body are inseparable. The body in the medial moment experiences its time in the circuit subjectively, even as it functions as an ensnared object. These are not two separate things, though it is difficult not to parse them in order to explain: the sensing subject and what it wants *is* the body and what it does. The subject is the charge traversing the wire, which is also the wire itself, a product of the wire's particular molecular structure.

Imagine, then, a predictive genealogy—a media teleology—of the soldier ensnared in the signaling circuit. Imagine the present and future communication technologies that would render him productive to the end, a machine that would make use of even the subjective impulses in his maintained mediality.

The soldier is alone in the desert. The soldier–subject is sick of signaling. The soldier–subject is in existential conflict with the soldier–medium, and is experiencing untenable angst. The soldier–subject awakens in a moment that compounds a vast collection of such moments that span over years, with a taste of adrenalin and bile. The soldier–subject is in his body, and it—the body—is dusty, nauseated, terrified, sleep-deprived, and deeply disenchanted. For a moment it perceives itself outside itself from the perspective of the all-powerful apparatus in which it is a mere atom of copper in a conductive conduit. *I am nothing,* the soldier–subject sees. *I am still nothing if I disappear. I was rendered this way, in fact, for the purpose of disappearing. And the me that is left that is not for this purpose is exhausted.* And angry. And despairing. Or, perhaps, the converse. The soldier–subject is experiencing a transcendence of himself as soldier–subject. He sees that he is nothing not with existential nausea but with the overwhelming joy of escape into that nothing, into one. A cause. A signal flash and oblivion.

Either way—whether charged with the desire for self-annihilation or the desire for transcendence of self—the soldier–subject is suicidal. He is insane. The grid of attributes that label him insane is thick and thorough and multilayered. The science that legitimates the statement that he is insane began centuries before this soldier–subject was born. It began in a confessional box,[60] where his ancestors were enculturated to share their evilest impulses in order to be cleansed. It continued on a therapy couch, where the cleansing became a dual-purposed mode of mental hygiene and data collection, where case study after case study produced a symptom-defined syndrome, finally "authorizing a relation only through the abstract universality of disease"[61] and thereby explicitly authorizing the eradication of the disease. The science also informed new kinds of contemporary confessionals: a public care for the self and a governmental care for the soldier fit for duty, a series of yearly physical and psychological screenings, a practice of talking about the troublesome symptoms of post-traumatic stress disorder and head trauma. The guys talking together after the unit's cautionary viewing of *Full Metal Jacket*. The enculturated, embodied, accepted understanding of military madness.

Perhaps the soldier is sporting a wearable Fitbit-type technology, issued to him by the Signal Corps, and perhaps the signal that it emits is a psychological "check-in" based on biofeedback. Perhaps he is posting messages on social media; his phone is counting the number of messages sent home and the odd hours at which they are sent.[62] Imagine advertising comfort to this subject in the way that a consumer-oriented circuit might advertise a cup of coffee or a vacation. The promising possibility of suicide haunts the soldier's media feed in suggested readings and banner headlines. It pops up like a pair of boots you looked at online and haven't bought yet. It masquerades as concern: "3 warning signs not to ignore," "If you or a friend is thinking of . . ." It isn't concern, though; it is actually a suggestion. A great data-sifting screen renders keywords and time stamps and heart rates and total hours of REM sleep into a signal. When the signal says "suicidal," the circuit quickly checks to make sure no friendlies are within range and ignites. The soldier becomes a bomb.

2

Soldiers in the Circuit

Media and Medicine in the First World War

> *Whatever democracy may be, it is supported by the mechanical processing of anonymous discourse.*
>
> Friedrich Kittler, *Literature, Media, Information Systems*

> *State is the name of the coldest of all cold monsters. It even lies coldly and this lie crawls out of its mouth: "I, the state, am the people."*
>
> Friedrich Nietzsche, *Thus Spoke Zarathustra: A Book for All and None*

Friedrich Kittler argues that "according to the conditions of 1890, all that matters is the technological ordering of all previous discourses."[1] For Kittler, the development of typewriters, both in the form of machine and (predominantly female) typists, is at the root of a grand, bureaucratic revolution and the (supposed) emancipatory trajectory of twentieth-century democracy. Here, then, Kittler's anonymous discourse is mechanically processed by female typists. On this point, we argue that Kittler has misdiagnosed the medium and the technology, for it was electromechanical tabulators and their use by female "computers" in the late nineteenth century that most clearly marks the emergence of the circuit that would become the most influential in the twentieth. Here we see the emergence of a governmental a priori in media and technological development. As John Durham Peters notes, "Numbers mediate and orchestrate distant complex social relations," and, more specifically, "Numbers in all their impersonality are democracy's ideal language, suited for gods, machines, and collectives."[2] Numbers are the cold language of the state and its statistics; they are a means of rendering or mediating aggregates, collectives, and complex systems for biopolitical and logistical governance.

Yet, numbers within the realm of governance are most often meant to index some*thing* or some process; citizens who live in cities, bushels of wheat, GDP, votes necessary to win the electoral college, the annual inflation rate, life expectancy of middle-class Caucasian women, how quickly Soviet ICBMs could reach North American cities. Numbers are the means by which risk is assessed and by which problematizations are legitimated. When a number is said to index an individual, as with a soldier, that number becomes not a manipulable element of mathematics but a means of addressing a file. The number is not an identity, but a mode of identification.[3] It produces addressability within a circuit. That address, that ID number, must be both mobile (as it follows a soldier to the front lines on another continent) and fixed (a permanent referent for the correct file). The ID number is referent to the file number when a simple accounting is mandated, but it also stands in for a nearly limitless amount of other numbers that exist in hosts of cross-referenced files. This chapter addresses in part how a complex set of governmental problematics are addressed through the circulation of the ID number as it travels with soldiers: first, as they are assessed by health and psychiatric officials from the mandate to create a national selective service at the outset of U.S. entry into World War I; second, to the movement of troops through the medical apparatus created by the military to treat wounded, ill, and psychologically traumatized soldiers; and third, to the dismissal of soldiers at the end of the war. Our story provides analysis of each of these three moments as they work together to push forward new circuits of governance and logics of assessment and optimization.

We suggest that the circulation of soldiers and the numerically rich files associated with them established a new set of protocols for governing the biopolitical health of the nation and for establishing a means for measuring the value of life according to the principles of triage. We broaden the notion of triage, originally developed by the French military to rank the relative need of injured soldiers, to account for various kinds of sorting mechanisms within the military machine, including selective service itself. Triage essentially mitigates the effects of short circuits, redirecting flows so as to minimize its effects and prevent an open circuit. From the circulation of potential soldiers to the circulation of postwar disability payments, we see the importance of information production (assessment), storage/access (file management and circulation), and information processing (ongoing recursive system-level assessment) as central mechanisms for managing populations in terms of security and risk. Whereas triage manages the circulation of damaged bodies at the front, selective service separates out the already damaged or insufficiently capable before unfit soldiers can arrive at the front lines where they might muck up the tactics of corporals and lieutenants.

The job of the surgeon general was to carry out a strategic initiative that accounted for the biopolitical health of the fighting force writ large, but also to minimize the damage that could be produced by over-enlisting soldiers from those defined as incapable. Managing the security of the fighting force meant minimizing the risk created by unfit vectors whether they manifested as casualties on the front lines or unaccounted-for "degenerates" undiagnosed by local draft boards. While triage may seem initially to be a more humane means for attending to the wounded, it begins to show itself as a mode of security maximization created through efficient circulation and screening of both bodies and numbers. Culling the population through selective service is simply biopolitical triage of the population. It is a solution that found one of its many potential problems. It is meant more to protect the military-industrial circuit than the citizen. Both numbers and soldiers must circulate in order for national military strength to be maximized. The Hollerith tabulating machine became a means by which the two forms of circulation were synchronized. People for the first time were turned into real-time data vectors.

While this is the broad story we want to tell, more immediately, we examine three uses of the Hollerith tabulating machine and the attendant technology of the "punch card" by the U.S. Medical Corps which operates under the auspices of the surgeon general. While this is often done via the maintenance of soldiers' and other personnel's health, it is also done via the use of contemporaneous medical doctrine to maximize the ability of the armed services to "selectively" choose which members of a population are appropriate to military "service."

Lastly, throughout this chapter we will be employing the terms that statisticians, doctors, bureaucrats, and the U.S. Army used to categorize, manage, and extract value and utility from human bodies. We understand, as David Mitchell and Sharon Snyder have argued, that "disability underwrites the cultural study of technology *writ large*."[4] Rather than coin new terms or employ scare quotes ad nauseam, we have relied on the actual language used in the archival documents we've examined, which, while problematic, helps to make clear the (ableist) normativity of the circuit we are analyzing here. While our analysis will focus more on the logic of this circuit and the ways in which it problematizes bodies in accordance with that logic, the ground is ripe for interventions from a disability studies perspective that we hope will be explored elsewhere, by us and others. The deep crosscurrents between disability studies and media studies are just recently beginning to garner the attention they deserve.[5] We hope this project can begin to show how the logic of the circuit automates and reifies ableist problematizations of the body.

Theory and Method

In this chapter we draw on the methodology of media genealogy outlined in chapter 6 and look to expand previous scholarship on media genealogy.[6] To do so, we posit a "governmental a priori" as an organizing force behind the development of media and communications circuits, particularly early computation. We position this in contrast to the standard military a priori that has dominated media studies scholarship, particularly archaeologies and genealogies of early computation that tend to begin their analysis with the introduction of electromechanical analog and digital computers to calculate ballistic trajectories in World War II.[7] We argue that an important prehistory to electronic digital computation occurs from the 1890s to the onset of World War II in which certain computational paradigms and infrastructures had already stabilized and inflected the development of computational technologies and the sets of practices through which they were incorporated into our everyday lives. These paradigms and sets of practices were largely organized around circuits of governmentality.

As Geoffrey Winthrop-Young has argued, Kittler developed his "war answer" or "military a priori" as a means for overcoming the perceived limitations of archaeological analysis in providing an explanation for why one episteme replaces another.[8] For Winthrop-Young, Kittler's methodological commitments already fused archaeology and genealogy:

> German media scientists such as Kittler evolved their own link between archeology and genealogy; that is, they fused war to its discursive effects by examining the mechanisms and technologies

> of inscription, physical disciplining, and surveillance that connect the two.[9]

For Kittler, the military a priori can effectively fuse archaeological to genealogical method in the analysis of media and communication technologies. As Winthrop-Young notes, "War serves as the prime techno-historical catalyst and hence as the explanatory backdrop of media evolution; media evolution, in turn, explains epistemological patterns and ruptures."[10] Winthrop-Young sees three other plausible means for addressing archaeology's limitations. First, Paul Virilio develops a historical argument that the military necessity for speed and accuracy demands technological replacements for the human.[11] Second, Keith Hoskin specifies the disciplinary mechanism attached to new forms of "learning to learn" by which humans automate and mechanize the learning process.[12] And third, James Beniger argues that a series of "control revolutions" have taken place in civilian arenas that demanded the programmability of the human to coordinate with machines.[13] What we ultimately find in Winthrop-Young's examination are other prevalent genealogical explanations for the centrality of the military in the development of media and communications technologies.

We hope to add to these accounts an articulation of the governmental a priori that we see similarly driving media and communications technologies, and particularly computation before World War II. We hope to show that there exists a dialectical movement between governmental problematizations and the implementation of circuits. New media open up a new range of human conduct to be known, analyzed, and problematized, which in turn creates new capacities for governmentality. These resulting problematizations hence call forth new media for "clarifying" the emergent problematic by means of higher levels of resolution, increased storage, and more robust computation. As such, media are central to the governmental power/knowledge dynamic and necessary for any governmental imaginary to emerge. In short, governmentality provides a specific motor for media escalation, which we articulate through our concept of the circuit.

We agree with Kittler that war heightens and intensifies media concerns. It compresses the temporal dimension, demanding quicker responses, while simultaneously expanding the spatial dimension, demanding technologically mediated command and control systems to span the globe. The need for faster and globally distributed data analysis, something approaching circuits that can command and control logistical and military operations across great distances in near real time, is a constant consideration in war, as has been shown in chapter 1. The demands of war amplify the risk of faulty information retrieval, storage, and computation, which strongly encourages the automation of previously human tasks and the institution of rigid chain-of-command structures for those duties yet to be automated. As such, they constitute rich experimental grounds for new media and communications technologies, as the chain of command produces a large body of disciplined subjects, the suspension of constitutional rights allows for forced compliance and reduces concerns over privacy, and the push toward automation creates ample funding streams for experiments with new technologies, particularly computation. What these new circuits promise in the end is less a global logistics than a global programmatic, where logistical analyses connect with both predictive analytics and command-and-control structures to operate in real time across the globe.

In this chapter we show how the military provided just such a fertile experimental ground for new governmental circuits that had less to do with traditional military concerns like battle strategies and tactics and more to do with the regulation of bodies and supplies through statistical aptitude tests, statistics-driven medical practices, and statistics-driven logistics. The points of overlap between these new governmental techniques and technologies were embodied in the figure of the Surgeon General of the Army, the Provost Marshal General of the Army, and the quartermaster general. In the following section we will examine the biopolitical problematization of managing soldiers' health and aptitude for combat that emerged during the Civil War and manifested most clearly in the attempts to manage Civil War pensions during Reconstruction. Next, we look at the introduction of Herman Hollerith's tabulating machines in 1890, which would later become the cornerstone models sold by International Business Machines (IBM). Following that, we examine the use of statistical and computational methods by the Surgeon General of the Army and the Provost Marshal General of the Army to facilitate the Selective Service Act, mobilize millions of men for World War I, manage their combat readiness and health throughout the conflict, and establish a bedrock of computed statistical knowledge after the war that would later be central to instituting the Social Security Act of 1935 and mobilizing troops for World War II. In closing, we look at the historical impact that this early conjuncture in governmental computation has had and examine how it will continue to influence future governmental research and development of computational technologies.

The Problematization of the Civil War

The library of the Surgeon General's office once sat on the Mall in Washington, D.C. The building used before that had previously been the Ford Theatre, the site of Abraham Lincoln's assassination. Two years after the emancipator's untimely demise, the library's primary function was to give life to a system that would provide the medical information necessary to sustain Civil War veterans' pension payments. All Union soldiers who volunteered or were conscripted to fight in the Civil War were promised a pension fund at the end of their service that could be collected by them or their next of kin in perpetuity. This was one of the first, and certainly for its time the largest and most sophisticated, arms of the federal bureaucracy, leveraging vast amounts of human labor, paper forms, and archival documents, proofs of identity, and computational labor in an attempt to distribute vast sums of capital to large numbers of people all over the country and over a long period of time. In our terms, this was one of the most sophisticated governmental circuits ever instituted. It is worth noting that at this time, computation was still a human task and the people who performed it were called "computers." This original use of the term persisted even as computation was increasingly automated by electromechanical tabulators, analog computers, and electronic digital computers, only being fully disambiguated into our contemporary usage of the term in the latter half of the twentieth century.[14]

As one might imagine, this first attempt by the federal government to maintain a registry of millions of people and account for and disburse their pension payments was fraught with a long series of public failures. By some apocryphal accounts it was in the Civil War pension program that the term "red tape" originated, based on the literal red tape used to bind the pension files whose incompletion or inaccurate completion often led to a failure to locate and compensate Civil War veterans and their dependents. Here the urgency to continually enlist fresh able-bodied soldiers to be mown down in combat with the Confederates carried aftershocks of war in the necessity to keep track of and monitor returned veterans and their next of kin in order to equitably distribute the medical and economic remuneration that had been promised them to induce them to enlist in the first place.

If we are looking to a point at which Kittler's war a priori combines with a governmental a priori—the distribution of risk and its management—via media escalation, then we might see this as a point at which addressability within governmental circulation becomes a point of crisis. The federal government needed to know the terms of address (names) for registered veterans and next of kin to verify their identity as well as their addresses (physical locations within the postal grid) in order to tender payments, and it needed a continually updated register of addressability so that it could continue to perform these functions on an ongoing basis. It also needed to maintain up-to-date records of veterans' health, applications, certificates, and military records. These data were kept on alphabetically organized pension cards that referenced numerically organized pension applications and certificates (see Figures 2.1 and 2.2). From the perspective of veterans and their families, the system failed frequently and publicly. This crisis of addressability would only be resolved with the introduction of punch-card tabulating equipment, birth certificates, and the assignation of numerical identifiers to its military members via "dog tags" in the early

NAME OF SOLDIER: Hurst, John W.

SERVICE: Late rank, Private Co. I, 110 Reg't Ohio Inf.

TERM OF SERVICE: Enlisted Feb. 22, 1864 Discharged June 25, 1865

DATE OF FILING.	CLASS.	APPLICATION NO.	LAW.	CERTIFICATE NO.
1876 Dec. 12	Invalid,	228607	T	151803
	Widow,			
	Minor,			

ADDITIONAL SERVICES:

REMARKS: See Moth. Orig. 846797 Aaron C. Hurst, F 18 U.S. Inf.
See M.C. 534641 Foster C. Wable, C. 142 Pa. Inf.

Died ..., 1..., at ...

Figure 2.1. Civil War pension card. National Archives.

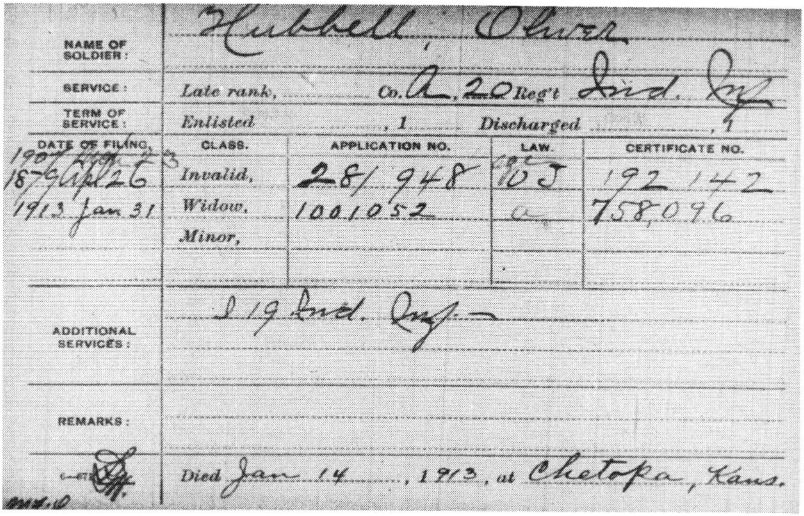
NAME OF SOLDIER: Hubbell, Oliver
SERVICE: Late rank, Co. A, 20 Reg't Ind. Inf.
TERM OF SERVICE: Enlisted , 1 Discharged , 1

DATE OF FILING.	CLASS.	APPLICATION NO.	LAW.	CERTIFICATE NO.
1879 Apl 26	Invalid,	281 948		192 142
1913 Jan 31	Widow,	1001052		758096
	Minor,			

ADDITIONAL SERVICES: I 19 Ind. Inf.
REMARKS:
Died Jan 14, 1913, at Chetopa, Kans.

Figure 2.2. Civil War pension card. National Archives.

twentieth century and later universally through the Social Security numbers necessitated by the Social Security Act of 1935.

From a governmental perspective, the federal experiments with the Civil War pension program produced a successful circuit. The records were collectively manipulated and computed to predict the cost of the program in perpetuity, a grim calculus of remuneration for spent bodies returned to an ableist society. Further, the tabulations of these data were also central to future military planning, particularly during the preliminary stages of mobilization for World War I. Dr. Benjamin A. Gould did extensive studies of soldier physique during demobilization in 1865, which he published in 1869 as *Investigations in the Military and Anthropological Statistics of American Soldiers* in two volumes.[15] Gould used what he termed an "andrometer" (see Figure 2.3) to partially automate the collection of anthropometric data and produced a data set that is still used by historians and anthropologists today to examine changes in things like American diet and physique since the nineteenth century.

These Civil War records provided the first materials required to produce data like that seen in Figure 2.3. In other words, a storable and addressable file that could be updated and collectively collated allowed for

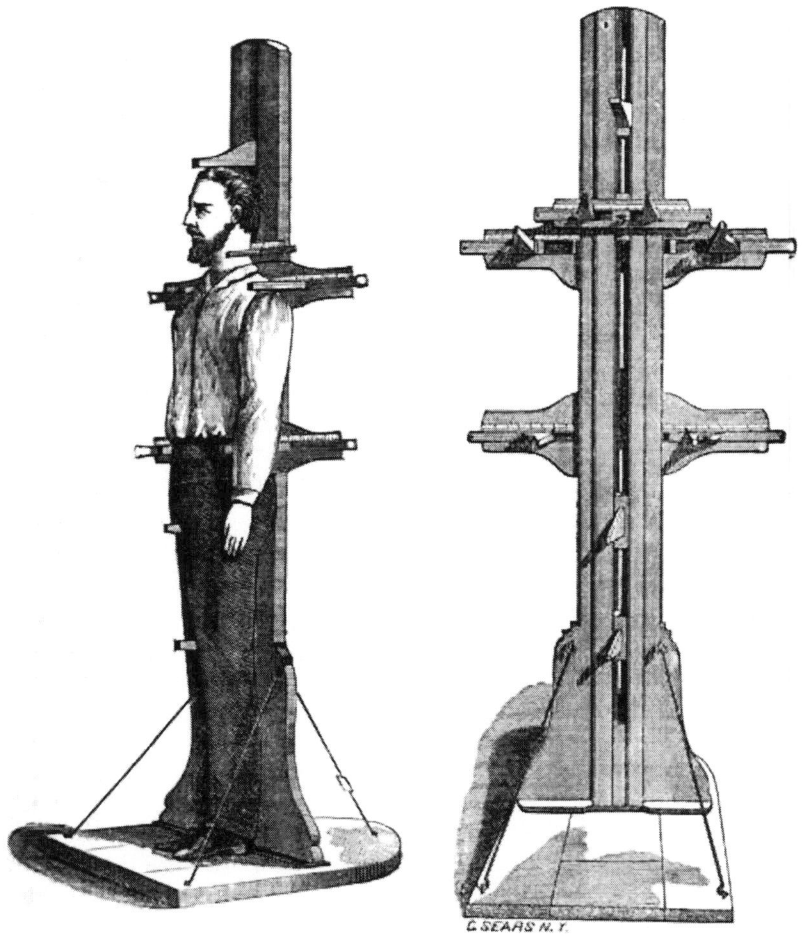

Figure 2.3. Dr. Gould's andrometer. Benjamin A. Gould, Investigations in the Military and Anthropological Statistics of American Soldiers *(Cambridge, Mass.: Riverside Press, 1869), 235.*

an extension and intensification of the federal government's temporal capacities—its ability to maintain control over time and via time management. The last widow of a Civil War veteran collected his pension until 2008, and as of 2017 there was still one child of a Civil War veteran collecting monthly pension payments.[16] For 150 years, this circuit guaranteed that pensions were delivered to Civil War veterans and their dependents.

Herman Hollerith and the Birth of Electric Tabulators

Before analyzing the governmental circuits put into place to mobilize and manage American bodies for World War I, it is worth examining the actual mechanism that was deployed to enable that circuit—the electric tabulator. Herman Hollerith, born on February 29, 1860, in Buffalo, New York, would go on to develop the first electric computation machine capable first of processing, and later of storing, vast quantities of data. He was educated at the Columbia College School of Mines, which he left in 1879 to take up work on the 1880 U.S. Census, though he would later submit essays on his patented work to receive a PhD in 1890. While at Columbia, Hollerith received a rigorous education blending elements of practice with theory, of physics and chemistry with engineering, though he never took a course on electricity or statistics.[17] Electrical engineering did not exist until 1882, after Thomas Edison's invention of the incandescent lamp and the installation of the first electric company, the Pearl Street electric power system, in New York. Classes in electrical engineering would not be offered at Columbia until 1889.[18] At the 1880 Census Office in Washington, Hollerith was assigned "to collect statistics on steam and water power used in iron and steelmaking," and also voluntarily computed life tables for Dr. John Shaw Billings, a special agent directing the vital statistics division of the census. It was Billings to whom Hollerith would grant full credit for the inspiration of his tabulating machine.[19]

While there are multiple accounts of the actual source of Hollerith's inspiration, nearly all cite the earlier Jacquard loom, despite different narratives for how Hollerith first encountered it. In 1804, Joseph-Marie Jacquard, a Lyons-based silk weaver by trade, patented a new loom that employed punch cards to automatically control the system. James Essinger notes that, at the time, the Jacquard loom "was unquestionably the most complex mechanism in the world."[20] It was automatically and continuously fed in a reliable and rapid way, and was exceedingly flexible. Employing as many as fifty thousand punch cards for a single pattern—most notably Jacquard's own face—these looms were capable of processing an extraordinary amount of data.[21] They have subsequently been the inspiration for the majority of the attempts to build the world's first computers, which Essinger refers to as "the first Jacquard looms that wove information,"[22] including both of Charles Babbage's engines and the Hollerith tabulating machine. It is unclear exactly how Hollerith came into contact with the looms, which were very popular at the time. One account has Dr. Walter F. Willcox, an associate of Billings's, quoting Billings as saying that he had specifically mentioned the Jacquard loom to Hollerith during their dealings at the 1880 census.[23] Other accounts trace Hollerith's first interaction with the Jacquard loom to his brother-in-law, Albert Meyer (not the Albert Myer discussed in chapter 1), who was in the silk-weaving business himself.[24] Before leaving the census, Hollerith took a position as a clerk to get a feel for the work of computation, and he left believing he had a potential solution to the government's data crunching problems.[25]

After spending about a year developing his ideas while teaching at MIT, in 1883 Hollerith left to take up a position as a clerk at a patent office. The average number of patents that were processed in the 1880s was 32,277 per year.[26] Many of them dealt with electrical engineering, which had just been introduced in 1882 in New York City. Hollerith pored over the patents and resigned just six months later to begin developing a prototype of his own machine. He was already harboring the idea that would lead to the first electrical processing of digital information, in which a hole in a punch card would stand for a numerical quantity or the presence or absence of a specific item or quality. This idea would lead to a series of thirty-one patents spanning the years 1889 to 1919.[27]

While Hollerith's initial prototype was a paper tape machine,[28] in which a continuous strip of paper contained a series of two holes corresponding to various pieces of data, he quickly shifted from paper tape to punch cards, which offered great durability and the capacity to be sorted for batch processing. While these punch cards were tailor made to individual applications, they most frequently contained twenty-four columns of quarter-inch squares, for a total of 288 data points, and had clipped corners to assist in stacking and sorting them.[29] The presence of a hole at any of these 288 data points would trigger an electric relay connected to a counting (or, later on, an adding) machine. Hollerith designed a series of punching devices to facilitate the tedious labor of inputting data onto punch cards by punching and verifying punched holes. These developments allowed for the expansion of punch cards to store more data and meet the growing demand of government and private industry for more sophisticated tabulations. Hollerith's punch cards increased from 24 to 36 to 45 to 80 columns, the last two becoming industry standards in 1907 and 1928, respectively.[30]

The tabulating machine itself was made up of a press connected by a series of electric relays to a series of counters. Punch cards were set into the press on top of a rubber plate that contained 288 holes corresponding to the positions in the card. Each hole contained a deposit of liquid mercury, which was connected by a wire to a binding post on the back of the supporting framework. Hinged above was a pin box, with 288 spring-loaded pins, again corresponding to the positions in the card, each grounded and connected to a barrel of battery acid to provide electricity. When pressed down, the pins corresponding to positions in the card with holes would enter the mercury and complete a circuit, while those corresponding to positions without holes would retreat into the spring box. The circuits initiated would activate electromagnets connected to dial-based counters divided into one hundred units, each with two hands, one for individual units, and the other, employing a carrying device, for hundreds of units. The counters could thus advance to 9,999 before needing to be copied down by hand and manually reset one at a time. There were forty counters to a machine, each three inches square, which could be plugged into the machine without needing to rewire anything.[31]

For cross-tabulation purposes, each tabulating machine was equipped with a sorting box, connected by electric relays to the press. The first sorting box was a horizontal model, with a series of compartments held closed against the tension of a spring. When a card was tabulated, an electric relay would trigger an electromagnet in the armature at the front of the box, which would cause the corresponding sorting compartment into which the card was to be placed to open automatically. The machine's operator then easily closed the lid after having deposited the card into the appropriate compartment. Cards could be sorted by individual data points, or by multiple data points in combinations with one another.[32] Hollerith later developed a mechanical feed and vertical sorting box to increase the speed of processing through automation. The mechanical feed required a new processing method, and Hollerith developed one similar to his earlier idea of the paper tape processor, which he called dynamic brush card reading. Here cards acted as insulators, passed between brass rollers with a series of steel brushes aligned with each column, which closed circuits each time they touched brass through a punched hole in the card. As Lars Heide notes, "All parts of the machine were synchronized with the movement of the card and registered the digit value of a hole according to its row on the card."[33] These machines were much more complex to manufacture, as the original 1890 machine had no turning parts, whereas the new machines contained electromotors; however, processing speeds were greatly increased, with average speeds of 210 cards per minute, about five times that of a manual feed. After processing, cards were dropped into one of thirteen chutes, twelve for sorting and one for rejects. The only drawback was the fast processing required frequent emptying of the sorting box, as each chute could hold only two hundred cards before it became overfull and began to crumple the punch cards.[34]

Perhaps the most impressive feature of the tabulating machine was its incorporation of electric relays to allow for programming. Relay circuits were capable of opening or closing all the relays involved in any given combination, at which point the counting circuit is completed, and its impulse goes through all of the relays remaining open and triggers the counting mechanisms' electromagnets. The use of relays allowed Hollerith to perform cross-tabulations at the same time as simple headcounts. While it is conceivable that Hollerith could have built machines containing enough relays to perform very advanced tabulations, in practice he was limited by the number of counters, the difficulty of programming and maintaining such a sophisticated wiring schema, and the ability to supplement for this loss through the sorting box. Thus, most machines ran with no more than three or four relays in any given circuit.[35] Programming the machines was very difficult work, as many small wires had to be disconnected and soldered into new positions for each tabulating process. Hollerith solved this problem by renting the machines out and guaranteeing their maintenance for the duration of their use.[36] The renting of tabulation machines established a precedent that would be kept up for many years afterward.

In order to compensate for the difficulty of programming, Hollerith added some features to his machines that he would perfect throughout the years in order to help their operators make simpler programming choices and detect errors. Hollerith devised a plugboard in the head of the press, which consisted of a series of sockets that would allow operators to shift the configurations of little cables in order to alter the programming of the machine during processing or between tabulations. He also connected bells to the counters that would ring each time a card was correctly registered or an out-of-place card had mistakenly been processed.[37] This advanced use of electrical circuitry actually allowed for a much simpler machine than its fully mechanical equivalent, and also offered a great deal of programming flexibility.[38]

Hollerith piloted his machine in the tabulation of mortality statistics for Baltimore in 1887 and New Jersey soon after, and then in the tabulation

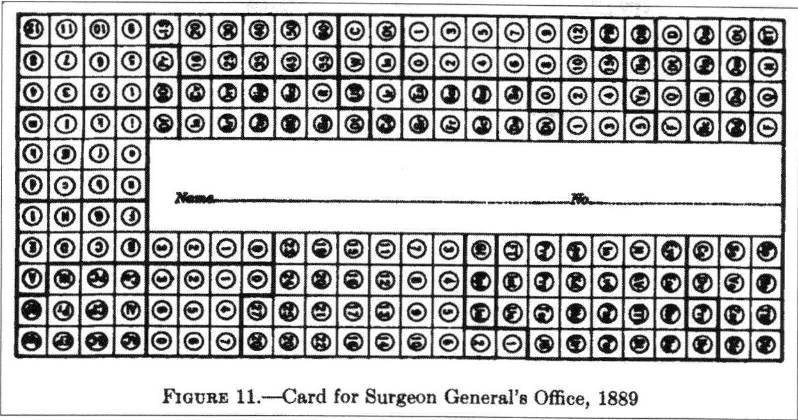

FIGURE 11.—Card for Surgeon General's Office, 1889

Figure 2.4. Punch card used in the Surgeon General of the Army's Office in 1889. Leon E. Truesdell, The Development of Punch Card Tabulation in the Bureau of the Census, 1890–1940, with Outlines of Actual Tabulation Programs *(Washington, D.C.: U.S. Government Printing Office, 1965), 39.*

of vital statistics for the New York Health Department in 1889.[39] In 1899 he also placed a tabulator in the Surgeon General of the Army's Office on a trial basis, the punch card for which can be seen in Figure 2.4. This relationship with the Surgeon General's Office would prove fortuitous, because, as we'll see, nearly thirty years later Hollerith's machines would be used to revolutionize military circulation. That same year Hollerith was granted the contract for the 1890 U.S. Census. By the end of the census, operators of his machines were each processing around seven to eight thousand cards per day.[40] The years after 1890 saw many other improvements in Hollerith's machines, though many of them were reserved for private industry. By 1907 the machine had crystallized into a rather stable state with dynamic brush card reading, thirty-six-column punch cards, keyboard punches, plugboard programming, automatic vertical sorting boxes, and compact adding machines (usually up to five per machine). By 1910, Hollerith's work had paid off; he could list many of the most prestigious and wealthy companies in the United States as his clients. In 1924 his company would be renamed the International Business Machines Corporation, which would enshrine Hollerith's computational standards for decades with mostly minor adjustments.

Despite playing a pivotal role in the crystallization of modern computing, punch-card electric tabulation is frequently under-analyzed. But, more important for our purposes, it also demonstrates some of our key conceptual distinctions about the circuit. As you've likely noticed above, the use of paper cards with discrete binary punches and the need for real-time processing and human intervention all blur the line between the analog and the digital. Further, the tabulator, itself a circuit, is only functional within broader circuits of human computers and governmental administrations. Yet its programmability and the increasing diversity of the data available to be punched demonstrate its difference from infrastructure. Punch-card machines were frequently shifted between agencies for temporary use and for specific projects; punch-card data were shared, siloed, reproduced, and repurposed, all with little standardization (even the number of data points per punch card was customizable with sufficient capital). Lastly, as we'll see in the following sections, the use of these tabulators during World War I demonstrates a marked difference between the circuit and the network. While circuits share in some network properties, such as the emergence of statistical patterns at large scale and robustness in the wake of node death, they are much more interconnected with human bodies, much more serial and hierarchical, and simultaneously allow for centralized command-and-control structures and localized triage.

The Provost Marshal General in World War I

Provost Marshal General Enoch H. Crowder was in charge of using the draft to create an army. Further, he saw its proper use as having far greater effect.

> I have used the term "Selective Service idea," because that is what it means to me. Having lived with it, having made it the master of my thoughts, of my activities, my hopes, it has become an idea with me, a principle, a great light by which I can sense the orderly growth of this great republic.[41]

And the Hollerith tabulating machines were at the center of this effort in two fundamental ways. First, their successful application in three censuses proved their value, both in the sense of producing more information and doing so at lesser expense to the government. The U.S. Naval Department had already ordered Hollerith machines in the 1890s, well before there was a specific need for their use in Selective Service. These machines, like so many of Hollerith's early tabulators, spent most of their time lying dormant in between periodic censuses and other large-scale data processing tasks. In essence, they were a hammer waiting for a nail. That nail would manifest itself in the Selective Service Act, which in turn was structured such that the data it collected were easily transferred to punch cards for mechanical analysis.

Second, as has been shown, these punch cards were put into service by the Surgeon General's Office very early on and that office quickly saw their ongoing utility during and after the war for assessing the dynamics of the fighting force—as can be seen in voluminous reports that were issued during the war and published as early as 1919. These data could be used not just to reconfigure how soldiers were selected and how their health and ability was managed, but also to fine tune the media instruments used to collect this kind of data in the future. This latter use will be explored in the next section that looks at the Surgeon General's Office during and after World War I, while the former use by the Provost Marshal General's office for mobilizing troops through the Selective Service Act of 1917 will be explored here.

Military conscription, appropriately renamed "selective service," draws upon a logic of "natural selection" and the emerging, ableist discourse of eugenics as a means to strengthen and purify the nation's fighting force. This is explicit in President Woodrow Wilson's own rhetoric, where he writes, "The time has come for a more perfect organization of our man power. The selective principle must be carried to its logical conclusion. We must make a complete inventory of the qualifications of all registrants."[42] As the Nazis would later use IBM machines to aid in their eugenic campaigns,[43] the U.S. military used punch-card logic and Hollerith tabulating machines to aid in their selection and circulation of soldiers. The eugenic sensibilities of "selective service" became apparent following the war when the same machines were used to analyze a voluminous amount of data from the war effort that would serve as the basis for much U.S. eugenics science in the 1920s and 1930s.

As President Wilson described it, the "machinery" for the execution of Selective Service registration was assembled with shocking rapidity, but it "left room for adjustment and improvement."[44] The two major problematic vectors were physical and psychological deficiencies. Each received extensive attention and several volumes of data were produced before, during, and after the Selective Service process was being carried out in order to prepare for, carry out, fine tune, and assess the degree to which it could efficiently be used to take a mass of ten million men between the ages of eighteen and forty-two and shape it into an effective military. Specifically, selective service needed to select out deficient subjects, determine which subjects' skills and labor were more important to industrial production or essential social administration than to military service, create ranked subsets to determine the temporal ordering of when they would enter military service, and assess the best military placement of those selected. This massive circuit connected most of the American public with standardized metrics of ability and the needs of the military and the war economy. It was the first large-scale, electrically automated, and quantitative enactment of ableist normativity in the world.

Selective Service registration began with mandatory self-assessments. All men between the ages of eighteen and thirty-five (shortly after extended to forty-two) had to fill out a questionnaire to be returned to their local draft board, upon which the board would produce a registration card to be kept by the draft board and a registration certificate to be kept by the registrant (see Figure 2.5). These approximately ten million men were then required by law to carry their registration certificates with them at all times prior to deployment and were explicitly required to "display the same whenever called upon by a police official or member of a Local or District Board to do the same."[45] The registration certificates thus worked as one of the earliest forms of national identification, indexed and verifiable by serial number. The registration cards that were kept by the draft boards were thus also indexed by serial numbers. These cards included each registrant's full name, permanent address, age and date of birth, race, citizenship status, occupation and place of employment details, and details for next of kin, which were supplemented by a registrar's report listing the registrant's height, build, eye color, hair color, and report on any obvious disabilities that would physically disqualify him from military service.[46] It is worth keeping in mind that registration cards were constructed with tabulation in mind, their data collection and transcription being structured to facilitate easy transfer to punch cards, a point we will return to in the next section. Serial numbers were always written in red ink, a scarlet number by which their Selective Service data was written onto their bodies, articulating how they would be seen and see themselves in the war effort in accordance with military and governmental norms of ability.

Upon returning their questionnaires, registrants had to show up at draft

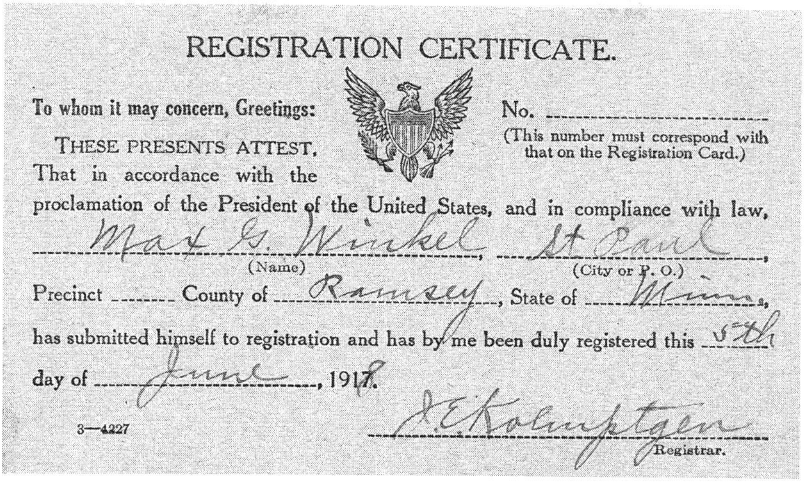

REGISTRATION CERTIFICATE.

To whom it may concern, Greetings:

No. ....

(This number must correspond with that on the Registration Card.)

THESE PRESENTS ATTEST, That in accordance with the proclamation of the President of the United States, and in compliance with law,

Max G. Winkel, St Paul,
(Name) (City or P. O.)

Precinct .... County of Ramsey, State of Minn.,

has submitted himself to registration and has by me been duly registered this 5th day of June, 1918.

3—4227

Registrar.

Figure 2.5. Selective Service registration certificate, 1917. National Archives.

boards not only to complete their registration cards and certificates but also to go through detailed questioning by volunteers working at the draft board and physical examinations by local doctors. The more detailed questioning included alphanumeric codes for professions as well as questions about the numbers of years on the job, salary information, previous military experience, educational background (including technical schooling), languages spoken, criminal records, whether or not registrants were "in sound health mentally and physically," with a request to indicate if they were blind, deaf, dumb, had lost a limb, were epileptic, paralytic, insane, or had a "withered or deformed limb," as well as questions about medical history, including doctors seen, treatments undergone, and institutionalizations in hospitals or asylums, and citizenship. It also contained special sections for registrants who worked for the government, were ministers of a religion, or were students in divinity or medical schools.[47]

Following registration, the men were subjected to physical examinations by local doctors who had volunteered to assist the draft boards. Physicians took measurements on and examined defects in the eyes and vision, ears and hearing, nose, mouth, throat, each individual tooth, weight, height, color, nationality, girth of chest (at nipples) at both expiration and inspiration, skin, "general appearance," hernia, genitourinary organs, anus and rectum, upper extremities, lower extremities, spine, and feet. They also took pulmonary data on the heart and lungs before, directly after, and two minutes after exercise. Lastly, they made notes on registrants' "mentality" (with a note to "exclude morons and imbeciles") and nervous systems (with a note to "exclude epilepsy").[48] After completing their examination, the doctors were charged with recommending the registrant be given one of the four following classifications:

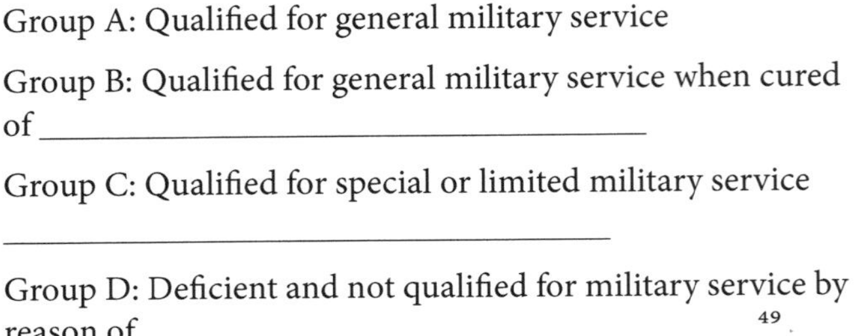

Group A: Qualified for general military service

Group B: Qualified for general military service when cured of ______________________________

Group C: Qualified for special or limited military service ______________________________

Group D: Deficient and not qualified for military service by reason of ______________________________.[49]

Once the recommendation was agreed upon by the draft board, and in some instances upon further physical examination by a medical advisory board, the registrant's classification was certified. These classifications were used to determine the temporal ordering in which non-exempt registrants would be mobilized and sent to the fronts.

These medical examinations were repeated by military doctors and volunteer psychologists at mobilization camps for all drafted men, creating a second normative data set of ability that could later be correlated to the data produced during Selective Service registration.[50] Of increasing importance among these tasks were screenings for "mental deficiency." Of the 442,275 men rejected from military service due to disability, 66,862 of them were rejected because they were considered unfit for military service due to intellectual, emotional, or moral defect or disease.[51] This number does not include the men whose conditions went undiagnosed or did not yet exist during mobilization who would later be hospitalized and discharged during the war, a huge portion of which suffered from what was called "war neuroses" or "shell shock." It similarly does not include the men who suffered from these conditions but were never discharged. Of all the men with diagnosed nervous and mental deficiencies, 12 percent were considered capable of military service. As Karl Bowman, a psychologist working with the U.S. Army during World War I noted, "It is thus evident that our army will, and must always, contain a considerable number of both actual and potential psychopathics, neurotics and defectives."[52] It is clear that the ableist norms of the military extended beyond the physical with little understanding of or respect for neurodiversity beyond the limits of its exploitability in combat and industrial production.

The goal is thus to secure the most effective fighting force possible by eliminating the mentally unfit as early and as accurately as possible. The fighting force will always still contain defectives, so the second task is to properly identify which defectives can still accomplish certain military tasks and circulate them appropriately. For example, Bowman argues that "manic depressives," people with "infective exhaustive psychoses," and those with "acute alcoholic conditions" can all engage in standard military duty at the front.[53] Bowman understood mental disease as "a failure or a difficulty of an individual to adapt himself to his environment."[54] As such, he makes a few concrete recommendations for how the military ought to respond. The first is to maintain universal military service so that the population will be more habituated to military duty from peacetime activities for eventual conflicts. The second is to better understand the different demands of military life in all of the branches and individual jobs of the military and combine that knowledge with examinations administered by psychologists to better determine which recruits are suited to which positions within which branches of the military.

Perhaps the greatest experimentation with the war was in identifying levels of mental defect and managing recruits' job placements and promotions within the military based on standardized psychological testing. Bowman defines mental deficiency as follows:

> Mental deficiency may be described as a lack of intellectual power, which is usually, but not always, congenital. The condition varies in degree from that of extreme idiocy to that of almost normal, and the gradations are so slight that it really grades insensibly into the normal. It is quite obvious in what way mental deficiency unfits for military service. The mental defective cannot understand or carry out properly anything but the simplest orders. He is very easily confused, lacks initiative, is easily frightened, and may cause or spread a panic.[55]

Because of these ableist and biologically essentialist interpretations of neurodiversity, the diagnoses of World War I were set to haunt the public in the ensuing eugenics movement in the United States. Throughout Selective Service registration and mobilization, 6.5 percent of all recruits were found to be mentally defective, and 24,514 of the 442,275 men rejected from military service were rejected based on mental defect.[56] These designations, considered heritable, would inflect government policy and catalyze scientific racism and ableism after the war. This emphasis on identifying mental deficiency also led to an alliance between the military and American psychologists, who looked on war mobilization as an opportunity to produce a large and standardized data set that might help transform their profession from an art into a science.[57] This effort was spearheaded by Robert Yerkes, who secured himself a position as colonel in the U.S. Army and oversaw the standardized testing of nearly 1.75 million recruits throughout the war.

Yerkes assembled America's leading psychologists, including W. V. Bingham, H. H. Goddard, T. H. Haines, L. M. Terman, F. L. Wells, and G. M. Whipple, most of whom had a markedly hereditarian interpretation of intelligence that would take an increasingly eugenic bend after the war, to produce the psychometric standards and write the Army mental tests later known as Army Alpha and Army Beta.[58] From there, Yerkes worked with the National Research Council to establish seventeen psychological committees to oversee the mental testing of recruits in thirty-five training camps and to analyze the results. The goal was for these tests "(a) to aid in segregating the mentally incompetent; (b) to classify men according to mental capacity; and (c) to assist in selecting competent men for promotions."[59] The Army Alpha test was designed for literate recruits, while Army Beta was a pictorial test designed for illiterate recruits. The test contains questions that required recruits to use basic mathematics and logic to unscramble sentences, find patterns, analyze analogies, fill in blanks, make common-sense-based deductions, and answer other basic word problems. The Army Beta tried to find visual representations for many of these same problems, as can be seen in Figure 2.6, where illiterate recruits were verbally instructed to indicate what was missing from each picture. After taking the tests, recruits were scored from A to E (with pluses and minuses) and recommended for certain placements within the military.

There were problems from the beginning with both of these tests. For one, they incorporated a lot of cultural bias that went unnoticed by their designers.[60] Take, for example, question 5 from Figure 2.6. What is missing here is a chimney, but recruits that grew up in areas where houses did not

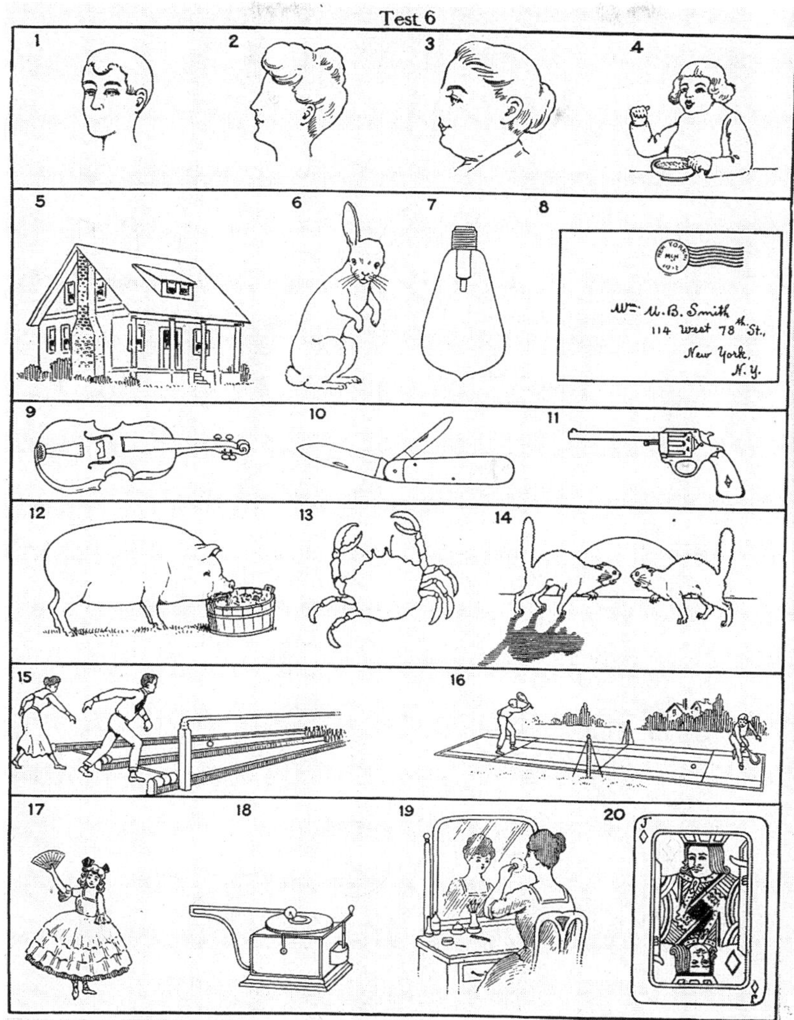

Figure 2.6. Sample Army Alpha/Beta test question. Division of Medicine and Science, National Museum of American History, Smithsonian Institution. The answers for this sample problem are as follows: (1) mouth; (2) eye; (3) nose; (4) hand; (5) chimney; (6) ear; (7) filament; (8) return address; (9) strings; (10) corkscrew; (11) trigger; (12) tail; (13) claw; (14) shadow; (15) ball; (16) net; (17) arm; (18) speaker; (19) arm in mirror; (20) diamond. There are a number of places online where one can take full versions of the original tests as well, such as https://openpsychometrics.org.

have chimneys may have had no concept that chimneys were necessary for a full representation of a house. These sorts of problems were compounded in the Beta Test because recruits needed to be able to see, hear, and understand the person giving the oral instructions for the test, on top of dealing with the anxiety of taking difficult tests in short amounts of time while facing pressure from both the Army and the psychologists. Stephen J. Gould has shown in great detail, as well, that the psychologists who administered the tests and later used the data for more public and eugenics-oriented arguments throughout the 1920s employed faulty statistical reasoning repeatedly.[61] However, this did not stop them from mounting a publicity campaign that would sediment their arguments into the Immigration Act of 1924, Jim Crow legislation, forced sterilization campaigns, and the SAT test.

Perhaps the most intense testing and data collection in all the military occurred in the Air Service, where screening for aptitude may have become more essential than in any other branch of the armed forces. As V. A. C. Henmon, a psychologist working with the U.S. Air Service to select apt pilots, notes, "We were constantly enjoined to remember that the flying officer was not to be an 'aerial chauffeur,' but a 'twentieth century cavalry officer mounted on Pegasus.'"[62] The severity of the task was not taken lightly. In addition to the Selective Service physical and mental examinations and the Army Alpha and Beta tests, prospective pilots were subjected to twenty-three aviation-specific mental and physiological tests—the most important measuring responsiveness to sudden excitation, perception of tilt, and mental alertness.[63] With the assistance of these new procedures, the Signal Corps, which had only fifty-two trained pilots in 1917, was able to grow into an Air Service with more than sixteen thousand pilots by the end of the war.[64]

As Eugene H. Crowder, the provost marshal general in charge of implementing the Selective Service registration, would note, "Within two months from the declaration of war a vast and complex machine had been erected, which, driven by a popular enthusiasm, enrolled a vast potential army in a single day."[65] In a remarkable passage, he describes the scope of the Selective Service registration's data collection:

> Never in the history of this or any other nation had a more valuable and comprehensive accumulation of data been assembled upon the physical, economic, industrial and racial condition of a people. . . . Scattered throughout the breadth of the land, lying in

> offices of the 4,648 Local Boards, 158 District Boards, 1,319 Medical Advisory Boards and 52 State Headquarters were the records of nearly twenty-four million men. . . . When it is considered that these records, as assembled, made of a file of cases which, if placed end to end, would make a line fifty miles long, and contained a volume of 400,000 cubic feet of paper, requiring approximately ten acres of floor space for their accommodation, the magnitude of the task that confronted the Department when it began their assembling can well be envisioned.[66]

For Crowder, the Selective Service represented a principle by which federal, state, and local governments could cooperate with citizens to integrate and distribute the processes of national governance, thus "preserving local self-government, yet making possible uniform, consistent, and efficient administration of national undertakings."[67] Crowder envisioned this circuit as offering new possibilities for managing class and interpersonal relations as well as federally administrating new domains like marriage and divorce, education, public health, and corrupt practices. While not all of these dreams would come to fruition, data collection on this scale and the automated data processing offered by tabulators after these data were transferred to punch cards did afford new biopolitical capacities for the state to govern public health, a paradigm that was perfected by the Surgeon General of the Army's Office during the war and standardized in tens of thousands of published pages in the years shortly thereafter.

Triage, Ambulances, and Surgeons General

Long after U.S. soldiers were selected and sorted according to their capacities, they were dispersed into an information-rich bureaucratic circuit that spanned oceans and continents and was constantly looking for new logistical efficiencies, predictive capacities, and informatic solutions to military problems. As much as soldiering is imagined to be organized according to the logics of combat, the maintenance of soldiers in terms of their health offers nearly as considerable challenges to military strategy and tactics. Caring for the wounded and limiting the damage of communicable disease were two of the most prominent challenges faced by militaries in general and those which fought World War I in particular. The U.S. military provides fertile ground for analyzing how the circulation of soldiers and the circulation of data accompanied each other as solutions to military strategic and tactical problems. We look to the joint processes of triage and the ambulance as mechanisms by which damaged bodies are integrated into the broad problematics of governing through circulation. Triage and the ambulance reconfigured the value/health of the individual soldier, but they also committed a very particular investment into the biopolitical health of the military and eventually of the nation. The circuit for health starts to be more directly mobilized as opposed to creating fixed spaces (clinics) around which the population would be organized. Rather than a spatial dynamic of health and non-health, the "dead zones" were brought to life through the use of the ambulance, which mobilized the immobile and drew them into the reach of care.

Medical treatises, scientific articles, military manuals, and government tracts were devoted to these topics. One such article provides a detailed account of the logic driving the systems that were developed:

> Picture a large open space surrounded by buildings. Into this there drives a motor ambulance. The tail curtains are opened and reveal four "lying down cases" on stretchers. These latter are swiftly and carefully slid out, and carried into a large receiving room 30 or 40ft. long. Another ambulance draws up with six or eight men who are 'sitting up' cases, and these are helped out and walk into the receiving room. The clothes of the patients are all thick with mud. Ambulance follows ambulance, for the field ambulances at the front have been filled up during the night, and there has been heavy fighting again at daybreak—a common hour for attacks—and thus it has happened that on many days from 500 to 1,000 or more wounded have arrived at a single clearing hospital in a single twenty-four hours. . . . But you must next appreciate that the hospital is only a "clearing hospital" or "station," and in its turn it must be promptly cleared of all cases that can be moved so as to be ready for next day's wounded. Therefore, ambulance trains must be ready daily to remove their hundreds to Boulogne or Rouen, or to hospital ships.[68]

All of this sorting and circulating takes place long after the wounded have already passed through a similarly elaborate circulatory system that includes stretchers, a "communicating trench," an Army surgeon, another stretcher or ambulance, a dressing station, and, before shipping off to the "clearing station," an anti-tetanus serum.[69] The trip to the "clearing station" is fraught, and routes ideally snake along "dead ground"—"ground which is not under the direct fire or observation of the enemy at the front."[70] Much of the logistical work of creating triage, ambulance, and field hospitals had been done by British, Canadian, Australian, and French forces well before the United States became involved in combat. However, once the Americans began operations, the circulation of the wounded was considered according to a spatiotemporal calculus that modulated according to the Hollerith tabulating machine protocols of data management. Speed was of the essence, but so too was the mitigation of the shelling of hospitals and the collection of ample data for immediate and long-term analysis.

The French military invented the ambulance to manage the counterproductive circulation of soldiers that accompanied the wounded from the field of battle. During the ongoing Revolutionary Wars of 1792–1802, the "Birth of the Ambulance"[71] was a response to pressures mounting from the democratization of soldiering that accompanied the "total wars" of revolution. These wars had broadened the fighting forces beyond the army of a king into an army of the people; this entailed a shift from a sovereign military to a biopolitical military that required prompt and expansive disciplinary training. Casualties rose significantly both in terms of real numbers and in terms of the percentage of soldiers taking part in battles. The mounting casualties that littered the fields of battle had to this point in European military history been left to suffer until the battle was finished, often for twenty-four to thirty-six hours. Movement of the wounded that did take place was seen as a misuse of the able-bodied who would haphazardly drag their compatriots *and themselves* away from the battle. It was noted by French military leaders that it might take as many as six able-bodied soldiers to move a single injured soldier. Such an inefficiency, and potential incitement to cowardice, was seen as a significant hindrance to military success. The creation of an ambulance (see Figure 2.7), as a contrivance of vehicles that moved surgeons to the front and moved wounded to a stationary field hospital away from the front, emerged to speed and monitor the flow of wounded and stanch the flow of healthy soldiers headed in the wrong direction. Once a soldier's ability status was altered within the grid of military demands during combat, his ensuing disability became a drain on the war effort that needed efficient governance.

The ambulance system, which would later be called a field hospital, demanded a mechanism for determining how to sort the wounded into

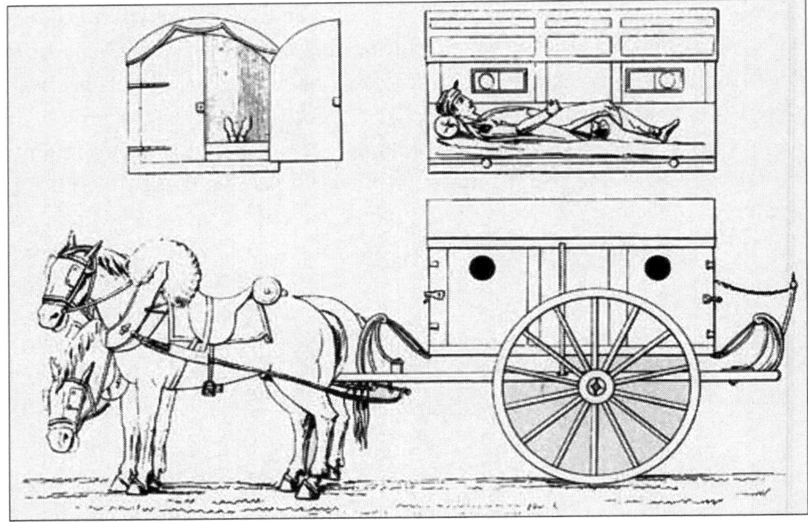

Figure 2.7. Early French ambulance design. Mémoires de chirurgie militaire, et campagnes du Baroin D.J. Larrey (Volume I) *(Paris: L'Imprimerie de J. II. Stone, 1812). Wellcome Collection.*

those worthy of care and those who should be left to suffer or die while fighting was still underway. The ambulance itself was organized to provide simultaneous surveillance of the soldiers and an optimal view of the tools of the trade. Pierre-François Briot, a French military surgeon at the time, boasted of the ambulances' effectiveness: "Supplies of rags, strips of all sizes, elastic sticking plasters, compresses of all types, bandages of all sorts; all was prepared, stacked, labelled, numbered in the boxes which never left the ambulance; everything was at our fingertips when we needed it. With one eye we saw the nature of the wound and with the other the most appropriate treatment."[72]

Triage developed alongside the ambulance. Triage was widely borrowed by U.S., Canadian, and British militaries from the French, who had been carrying it out in various forms since the early Napoleonic military campaigns. Historian of medicine George Thompson explains that "the sorting, classification and distributing done at a Triage required a skilled team to determine who was transportable and who needed to be retained until they were ready to be moved. Ideally the team had a thorough knowledge of medicine, surgery and human nature. Their evaluations had to be complete and unhurried but quick enough to prevent congestion caused by the arrival of new patients."[73] Again, we see that triage was a mechanism of sorting, but also, and more fundamentally, a means of ensuring the proper flow of military tactics on the battlefield. Injured soldiers are an undesirable blockage. They clot the arteries and veins of a healthy fighting unit. Their presence must be negated, but in a fashion that both accounts for their still-existent potential to become productive and the ongoing difficulty that their impeded circulation will produce for the military system. Triage turns the soldier into a case and necessitated a five-dimensional data-problematic:

> Listing them in the order of their priority, these steps were (1) recording the receipts of the reports, (2) examining the reports as submitted, (3) filing of the cards, (4) preparing the code, and (5) drawing up of a statistical card for the tabulating machines.[74]

This problematic was solved by the U.S. military through the adaptation of British models and applying them to the protocols of the Hollerith tabulation machine. Various forms of standardized documents first turned the data into what Lisa Gitelman has referred to as "paper knowledge,"[75] which through the punch card would become electrified and digitized, making it eminently more malleable, transferable, and, most important, *computable.*

Simple paper forms and envelopes were printed that allowed for standardized assessment, which would speed up the initial assessment process, simplify continued monitoring, and allow for easy transition to digital form. Taken in its entirety, this process constituted a brief but continuous record of medical care, ensuring the continuity of treatment and the identification of the patient at all times. Not until the patient was transferred to the United States and arrived at a hospital where definitive treatment was to be given was the record detached. The name of the hospital receiving the patient and the date of arrival were stamped upon the card, which was then forwarded to the Surgeon General's Office, where it was the one and only record in the War Department of the address of the patient that had been returned to this country for treatment.[76]

This data trail was an active circuit. While the end point might seem to be a U.S.-based hospital (Civil War pension payments suggest otherwise), throughout the process the raw data were being circulated for the production of periodic reports. Productively circulating the wounded through a secondary circuit that didn't short-circuit the primary one responsible for moving able-bodied soldiers to and from the front was the most important task of the medical services. But a third circuit needed to be developed that moved data in such a way that it accompanied and indexed an actual damaged body. The physical sorting of bodies was also organized according to schemata that were homologous to the triage process. The movement of bodies to and from the front was organized by the ableist normativity of the military's problematization of the body, and those designations were literally fastened to the body as it circulated within the war movement.

The outlines for such a system were present prior to World War I and were given priority in medical service branches. In an unremarkable speech given to the Canadian Military Institute in Toronto on March 6, 1896,[77] Lieutenant Colonel George Sterling Ryerson, a surgeon, provided a historical account of the development and continued implementation of the medical circuit that sorted, assessed, and cleared out wounded soldiers during times of war. Ryerson was quite keen to compare military successes in this development with advances in the electric circuitry coming to define the nineteenth century: "Military surgery has kept pace with the scientific advance of the century, and the field surgery of to-day differs as greatly from the septic scenes of horror of the sixteenth century as the telegraph does from the pony express."[78] Two overlapping circuits were developing for the logistical application of military medicine: triage and information management. The American Expeditionary Forces (AEF) on the Western Front would create a system that exhibited precisely these logics.

As Ryerson notes, "Previous to that time [the wars of Marlborough] soldiers who were so seriously maimed as to be rendered ineffective, were simply discharged, the State believing that it was cheaper to hire whole men than to restore the sick and maimed to health."[79] However, a change occurred in which the scientific application of medical care was programmed into the battlefield as part of the preparation for warfare. A calculus had to be drawn to account for a revaluing of the soldier that accompanied the updated capacity to heal. The investment in the population of soldiers was part political economy, part logistics, and part medicine. A codependent circulation of wounded and data needed to be created to assess, monitor, direct, guide, and fine-tune the biopolitical capacity of the fighting force. Much as Foucault describes other hospitals as places for care and surveillance, the various "triage stations," "clearing hospitals," or "lines of assistance" developed to triage wounded soldiers served multiple epistemological, tactical, and medical goals.

In 1812 the British devised the Royal Wagons Corps, their version of the French ambulance, which was made up of "special waggons with springs being constructed for the conveyance of sick and wounded."[80] Between 1812 and 1900 the British devised ever more tightly organized means for managing the processing and treatment of wounded soldiers, primarily organized according to an army corps—composed of forty thousand soldiers, the majority of which were sorted into three divisions of ten thousand soldiers each. Within any given corps "there would be 8 bearer companies, 10 field hospitals, 2 station hospitals, and 2 general hospitals, the latter being on the *line of communication at any distance up to 100 miles from the front.*"[81] Importantly, the whole enterprise would be guided by a surgeon general who had the most central role to "direct the measures for keeping the men in health, which is the main business of the army surgeon, so that at the critical time they be available."[82] In very simple terms we see here a spatiotemporal ordering of managing biopolitical capacity from the broad strategic state-level priorities, down and through the military rank and file into the management of the daily health of soldiers. Yet, binding the biopolitical to the national-strategic demanded modes of data collection, storing, and processing that speak to the specificity of a given media system and spatial dynamic. The United States followed suit in World War I and created schematics of an idealized system as well as maps of their actual distribution (see Figures 2.8 and 2.9).

Just as the medical units had to sort and organize the movement of casualties from the front through the various field hospitals, ambulance routes, and hospitals, their accompanying records and data had to be as carefully mapped, as it was the record, not the patient, that was referenced when determinations were made for how to handle the broken bodies and minds that were being cared for. This implied not only a spatial logic of mobility and distribution but a temporal component whose primary initial motivation was speed, whose long-term success depended upon sustainable storage, and whose assessment could only be accomplished via vast amounts of computation.

In the field of battle a very rationalized and programmatic form of turning wounded into information takes place. Ryerson describes seven "lines of assistance" that both provide care and act as nodes in the circulation of the wounded through a closed loop. Each line makes determinations regarding how to further process wounded soldiers and how to manage them as data; by "producing the [body] as a field of management," soldiers

Figure 2.8. Map of the 551 stations from which the Red Cross rendered service in France in World War I. Henry P. Davison, The Work of the American Red Cross during the War *(Washington, D.C.: American Red Cross, 1919).*

were becoming *biosubjects,* that is, their very biology became a locus of political power relations through attachment to technologies of identification, surveillance, selection, and maintained (bio)governance.[83] The first "line of assistance" provides medical supplies carried by mule, and stretcher bearers who can rapidly move wounded. This "line" also begins the informationalization process of wounded as they are brought to a "collecting station" where it is determined whether they go back to the fray or whether they are provided informational being through a "tally on which is stated the man's name, number, rank, regiment, wound, treatment, and any special instructions" which will accompany them as they are moved onto the second line of assistance.[84] The AEF created a "diagnosis tag" that collected this preliminary treatment data (see Figure 2.10).

The second line provides another clear point of demarcation as they arrive "on the right the severely wounded, and on the left the slightly

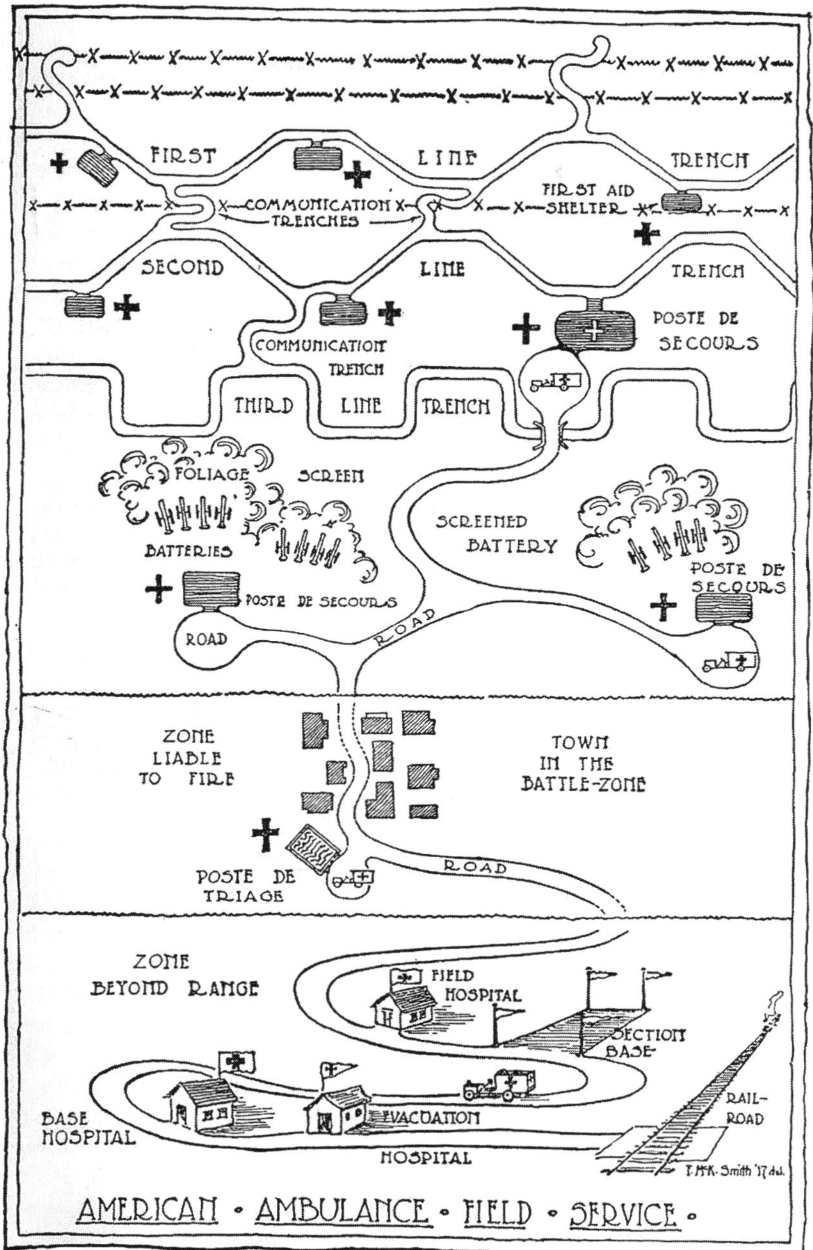

Figure 2.9. Chart of typical American ambulance service at the front in France. History of the American Field Service in France, "Friends of France" 1914–1917, Told by Its Members, Volume 1 *(Boston: American Field Service, 1921), 26.*

DIAGNOSIS TAG

FRONT

Date, hour and station where tagged: Oct.10, 1918, 6 P.M.; Dressing Station No.10

Name Roe, Thomas A.

Rank and Regt. or Corps: Pvt. Co. A, 16th Infantry.

Diagnosis:

G.S.W. left femur and right humerus.

Treatment:

Morphine sulphate, .016.

A.T.S. 1500 units.

Signature:

T.L.K. Capt. M.C.

BACK

SUPPLEMENTAL RECORD:

Primary hemorrhage, left thigh, moderate. Pressure applied temporarily over femoral artery. First aid dressing.

FIGURE 8.—Diagnosis tag used by AEF.

Figure 2.10. Field diagnosis tag. Albert G. Love, Eugene L. Hamilton, and Ida L. Hellman, Tabulating Equipment and Army Medical Statistics *(Washington, D.C.: Office of the Surgeon General Department of the Army, 1958), 57.*

FIELD MEDICAL CARD

FRONT

ORIGINAL SIZE 6"X9¾"

U.S.A. Transport Madawaska.

Date of entry, Feb. 26, 1919.

No material change in patient's condition while on Transport.

March 6, 1919. Transferred to Port of Embarkation, Hoboken, N.J.

P. H. N.
Capt. M. C.

Approved Abbreviations

Morphia .... M.
Anti-Tetanic Serum .... A.T.S.
Gunshot Wound .... G.S.W.
Fracture, Simple .... F.S.
Fracture, Compound .... F.C.
Fracture, Comp. Comminuted .... F.C.C.
Not Yet Diagnosed or Undetermined .... N.Y.D.
Fever of Unknown Origin .... F.U.O.
Disordered Action of Heart .... D.A.H.
Slight .... O
Severe .... S

(Stamped)

March 7, 1919. Received this date at Base Hospital, Camp Merritt, N.J.
March 8, 1919. Transferred to Walter Reed General Hospital, Washington, D.C.

L. F. J.
Capt. M. C.

(Stamped)

March 8, 1919. Walter Reed General Hospital, Washington, D.C. Received this date.

R. S. T.
Capt. M. C.

FIELD MEDICAL CARD

Name ... Roe. ... Thomas A.
(Block letters)

Rank ... Pvt. ... No. 2343276

Regiment or Staff Corps 18th Inf., Co. A.

Line of Duty—Yes

Field Hosp. No. ... Evac. Hosp. No. 26

Dressing Sta. No. ... Camp Hosp. No. ...

Date of Admission

Morphia
Dose and Time .016; 6 P.M., Oct. 10, 1918.

A. T. Serum
Dose and Date | 1500 units; 6 P.M., Oct. 10, 1918.

Diagnosis

(Over)

Base Hospital No. ----

Diagnosis (if altered from above)

Base Hospital No. ----

Diagnosis (if altered from above)

FIGURE 9.—Field Medical Card used by AEF (front).

Figure 2.11. Field medical card. Albert G. Love, Eugene L. Hamilton, and Ida L. Hellman, Tabulating Equipment and Army Medical Statistics *(Washington, D.C.: Office of the Surgeon General Department of the Army, 1958), 57.*

wounded."[85] From here the severely wounded are moved to the third line of assistance, the field hospital, which would complete a field medical card detailing the diagnosis and treatment of wounded soldiers (see Figure 2.11). These cards were placed into an envelope (see Figure 2.12) that would follow the wounded soldier throughout his circulation among military hospitals and treatment regimes, while the data would be computed in aggregate and communicated back to the Surgeon General of the Army's Office to compute periodic statistical analyses of troop strength, field hospital performance, and medical supply levels.

These mobile hospitals were strategically placed beyond the reach of enemy artillery, and their status as hospitals was originally conveyed to the enemy through a red cross during the day and two white lights surrounding a red light at night. However, after the war began and hospitals became targets, the placement of triage stations, clearing hospitals, and field hospitals were all concealed and subject to detailed evacuation plans

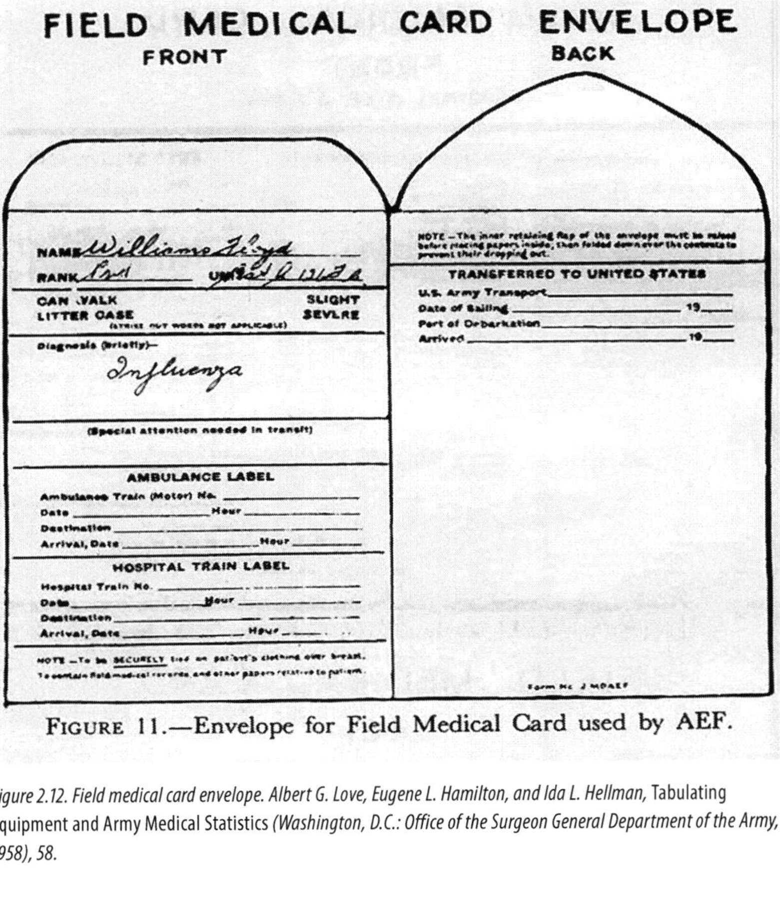

FIELD MEDICAL CARD ENVELOPE

FRONT

NAME

RANK

CAN WALK
LITTER CASE
SLIGHT
SEVERE

Diagnosis (briefly)—

Influenza

(Special attention needed in transit)

AMBULANCE LABEL

Ambulance Train (Motor) No.
Date
Hour
Destination
Arrival, Date
Hour

HOSPITAL TRAIN LABEL

Hospital Train No.
Date
Hour
Destination
Arrival, Date
Hour

BACK

TRANSFERRED TO UNITED STATES

U.S. Army Transport
Date of Sailing 19
Port of Debarkation
Arrived 19

FIGURE 11.—Envelope for Field Medical Card used by AEF.

Figure 2.12. Field medical card envelope. Albert G. Love, Eugene L. Hamilton, and Ida L. Hellman, Tabulating Equipment and Army Medical Statistics *(Washington, D.C.: Office of the Surgeon General Department of the Army, 1958), 58.*

to increase their mobility upon discovery and thus avoid targeting by the enemy (see Figure 2.13). The fourth line of assistance was a stationary field hospital farther from the front, and the fifth was a general hospital with four hundred beds. For the wounded too damaged to be redressed or sent back to the front, hospital ships, the sixth line of assistance, transported the wounded from various battlefronts back to hospitals and treatment facilities in England or the United States, like the Royal Victoria Hospital, whose duty was to "'clear the front of wounded men' who impede the movement of the army."[86] Taken as a whole, these lines of assistance work

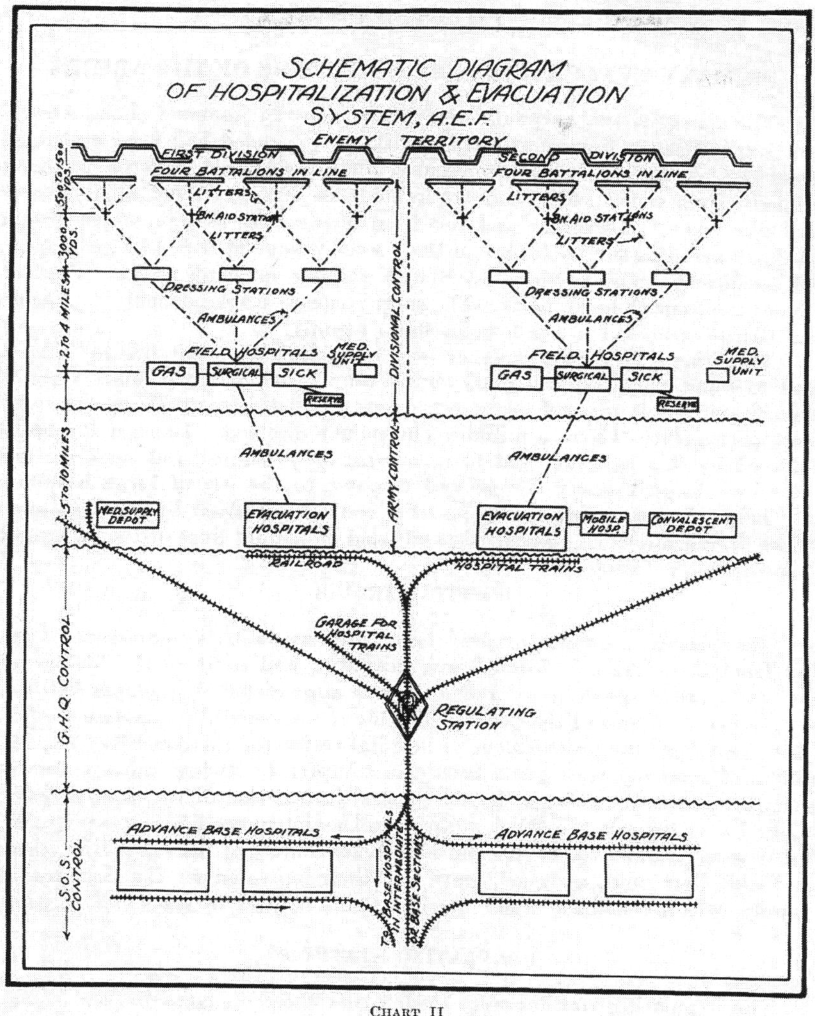

CHART II

Figure 2.13. Evacuation plan, part 2. The Medical Department of the United States Army in the World War (Volume VIII): Field Operations *(Washington, D.C.: U.S. Government Printing Office, 1925), 262.*

as an apparatus or "screening technology" for circulating the wounded out of military life or back into the military machinery, all the while producing a machine-readable data trail affixed to individual soldiers' bodies.

As mentioned, such general assessments of how the field hospital should work were largely put into place by the British (and copied by the Americans), who fittingly placed the transport of wounded soldiers into the portfolio of the Inspector General of Communications.[87] The terms used by the British for this developing apparatus were casualty clearing stations or clearing hospitals. While partly driven by a consideration of soldiers' well-being at the strategic level, the tactical concern was more direct: "Men should not be retained for treatment in the front-line units, where their presence might create difficulties and might interfere with the movements of the army. The prime duty of the casualty clearing stations was therefore to arrange for the satisfactory evacuation of the wounded and sick rather than their active treatment."[88] Wounded and sick soldiers blocked circulation. As such they needed to be removed through a separate, though interpenetrating, circuit. While various militaries borrowed tactics of the ambulance, triage, and field hospitals, as we'll see in the next section, it was the U.S. military that brought the electromechanical circuitry of the Hollerith machine as a mechanism for tracking and assessing the health of the fighting forces and the efficiency of the circulation system for clearing the wounded from the front.

U.S. Army Medical Statistics

For 150 years, managing the health of the United States has been the job of the surgeon general—quite literally a military general whose role is to administer medical knowledge and technologies to maximize the destructive and tactical capacity of the U.S. military. Though it may no longer be the case for the Surgeon General of the United States, there remain Surgeons General of the Army, Navy, and Air Force for whom this is still their mandate. National health or public health, insofar as such a thing existed, was first managed by the military and was treated as a military concern going back at least to 1798, when Congress established the Marine Hospital Fund, later the Marine Hospital Service, and ultimately today's U.S. Public Health Service. Militaries are particularly proficient at data on human subjects, as suggested here:

> The function of medical records is twofold, to provide for the care of the individual and to yield statistical information which will guide administrative policy leading to improved health conditions. Both of these principles are well known in the medical records of civilian hospitals and in the medical statistics collected by health departments. But in neither of these areas do we have an opportunity for these functions to be as fully exploited as in the military services where the population served is known, the medical services are more uniformly available, and the various medical records of the individual can be brought together.[89]

Here we'd like to focus on the military management of public health through the production and maintenance of medical records, much like those initiated with the Civil War pension, that was conducted out of the Surgeon General of the Army's Office.

As was already evidenced in the language of the Selective Service Act and the calls made by President Wilson for U.S. citizens to comply with registration even if they did not consider themselves fit for duty, the goals of the Selective Service System always extended beyond the mere mobilization of troops to send to the front. As Wilson noted in his call for a third Selective Service registration:

> Only a portion of those who register will be called upon to bear arms. . . . But all must be registered in order that the selection for military service may be made intelligently and with full information. This will be our final demonstration of loyalty, democracy, the will to win, our solemn notice to all the world that we stand absolutely together in a common resolution and purpose.[90]

Military service in World War I was an aid to larger progressive agendas, like building a national registry, producing verifiable identification for citizens, and collecting data for the rapidly expanding bureaucracies of the federal government for the ongoing, centralized management of the country's health, military power, and productive capacity. The volume of Selective Service data was greatly increased during the war by the Surgeon General of the Army's Office, which collected individual and aggregated accounts of all casualties and medical treatments received by U.S. troops throughout the war that could be cross-referenced with the Selective Service data.

As we saw in the previous section, the AEF collected detailed medical records that were literally and figuratively fastened to wounded soldiers' bodies as they moved through the military medical circuit from the front to field hospitals to general hospitals to domestic hospitals. Each AEF form was designed for easy transfer onto punch cards for tabulation during the war. A single punch card could track a soldier's medical diagnosis and treatment from the front in France to a domestic hospital in the United States (see Figure 2.14). All of the field medical records were transferred onto punch cards for tabulation during the war. Twenty-five women "computers" from New York who had worked with tabulators on the census and other endeavors were hired to work out of the Surgeon General of the Army's Office for the task. A further group of women was hired and trained for punching the cards. This repurposing of specialized modes of media labor and knowledge was essential to the success of the military's computational enterprise. Female telephone operators were also highly sought after by the military during World War I, and the U.S. military's telephonic expertise was a central part of their military media strategy.[91]

What we can see already in this attempt to produce increasingly real-time logistical and medical data for the whole of the armed forces is the extension of the governmental a priori, already solidified in the domestic arena through the Progressive movement, into the military domain. The semi-autonomy of military commanders necessitated by lag times in communication and data processing was becoming increasingly unnecessary within a governmental apparatus that had already successfully utilized telegraphy, telephony, radio, and punch-card tabulation to centralize the governance of everyday life in Washington. The new military would be run from an increasingly bureaucratic modus operandi that focused on proper collection, processing, storage, and transmission of clean, machine-readable data whose logical end would be centralized command and control through real-time satellite uplinked data streams—envision here the president and Joint Chiefs of Staff in the Situation Room with live satellite imaging and uplinked video from soldiers' helmets positioned between streams of population-level military data on troop strength and mobility.

This image was already in embryo in the First World War but was not fully realizable at the time. Data were still stored on paper and needed to

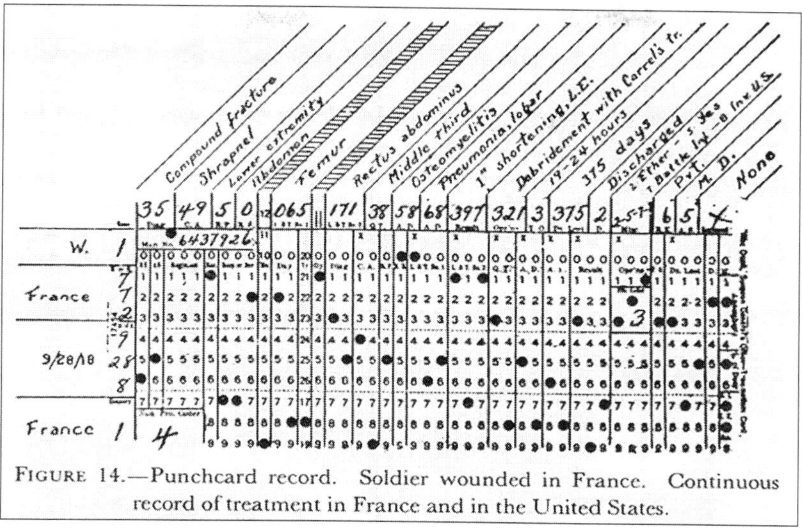

FIGURE 14.—Punchcard record. Soldier wounded in France. Continuous record of treatment in France and in the United States.

Figure 2.14. Punch-card record of soldier wounded in France's treatment in France and United States. Albert G. Love, Eugene L. Hamilton, and Ida L. Hellman, Tabulating Equipment and Army Medical Statistics *(Washington, D.C.: Office of the Surgeon General Department of the Army, 1958), 61.*

be transmitted physically before being hand encoded onto punch cards for tabulation. As such, the goal of daily reports of aggregate data were unfeasible throughout the war, and instead the surgeon general had to suffice with monthly reports on aggregate data that were used throughout the war to make strategic and logistical military decisions. These data were subjected to further statistical analysis during the war and published in the annual reports generated by the Surgeon General of the Army for the Secretary of War.[92] Its most enduring form, however, would be its use in a fifteen-volume analysis of the role of the U.S. Army's Medical Department in the war that was ordered by Major General William Gorgas, surgeon general from 1914 to 1918, and compiled and published between 1921 and 1929 under Major General Merritte W. Ireland, surgeon general from 1918 to 1931.[93] Across tens of thousands of written pages, charts, graphs, tables, and illustrations, the operations of the Medical Department were parsed in their minutest detail so that future government projects might learn from and improve upon their methods, extending an ableist normativity determined by military and industrial needs.

MEASUREMENT CARD FOR CLOTHING PATTERNS
DEMOBILIZATION–1919

Name John Doe Army Serial No. 278659 Home State New York

Organization Hdqrs Co., 313 Infantry Age 22 Color (Check in squares.) White ☒ ½ black ☐ ¼ black ☐ Negro ☐ ½ black ☐ Indian ☐ Chinese ☐ Japanese ☐ Other ☐ (Name.)

Place of observation Camp Dix Date of observation Sept. 10, 1918 Initials of officer in charge A.M.S.

Place of birth of—

	Country.	State or Province.	City or Town.
Self	United States	New York	New York City
Father	United States	New York	New York City
Mother	United States	New York	New York City

Nationality of father's father American Nationality of mother's father American

Nationality of father's mother American Nationality of mother's mother American

Native language of mother English

Religion of father Protestant

Hair, color (Check in squares.) Flaxen ☐ Dark brown ☒ Light brown ☐ Clear red ☐ Medium brown ☐ Red and black ☐

Other noteworthy racial traits

Eye, color (Check in squares.) Clear blue ☐ Light brown ☐ Blue with brown spots ☐ Dark brown ☒

MEASUREMENT CARD FOR CLOTHING PATTERNS

MEASUREMENTS—ALL METRIC

0. Weight	145.0
1. Height, standing (stature)	160.1
2. Span (maximum, between finger and tips of outstretched arms)	161.0
3. Height, sitting	86.0
4. Height of sternal notch	132.0
5. Height of pubis	76.8
6. Transverse diam. of shoulders at level of head of humeri	42.7
7. Transverse diam. of hips, level of crests of ilia	29.8
8. Transverse diam. of chest at level of nipples, arms elevated and flexed	28.8
9. Ant.-post. diam. chest; level of nipples	21.0
10. Second dorsal vertebra to styloid process of ulna (elbow bent, horizontal)	68.0
11. Circumference of neck, level of laryngeal prominence, perpendicular to axis of neck	38.0
12. Circumference of chest, level of nipples	88.7
13. Circumference of waist, level of umbilicus	83.0
14. Circumference of thigh (maximum)	55.2
15. Circumference of leg, just above patella	37.0
16. Circumference of knee, level of patella	35.0
17. Circumference of leg, just below level of tuberosity of tibia	31.9
18. Circumference of calf (maximum)	34.7
19. Inside length of leg, from gluteal fold to tip of internal malleolus of tibia	64.0
20. Size of shoe worn if fitted since July 15, 1919, under par. 14, S. R. 28.	8-C
Number of measurer	6
21. Height of knee	43.4
22. Length of forearm	23.0

DOTTED LINES INDICATE DIAMETERS
FULL LINES INDICATE CIRCUMFERENCES
NOS. 2, 3, AND 10 NOT SHOWN ON FIGURE

PLATE II.

Figure 2.15. Measurement card for new uniforms at demobilization. Charles B. Davenport and Albert G. Love, The Medical Department of the United States Army in the World War (Volume XV): Statistics, Part One: Army Anthropology *(Washington, D.C.: U.S. Government Printing Office, 1921), 41.*

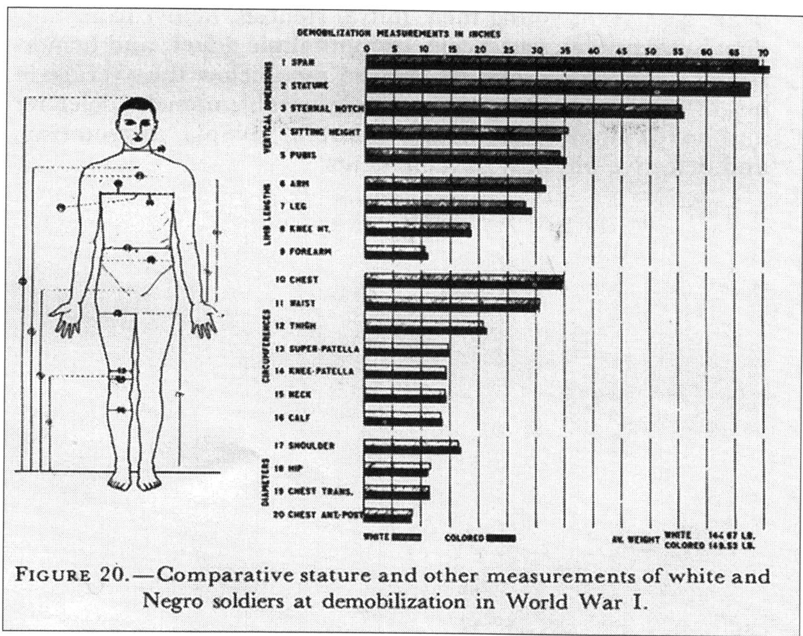

FIGURE 20.—Comparative stature and other measurements of white and Negro soldiers at demobilization in World War I.

Figure 2.16. Comparative measurements of 100,000 troops at demobilization. Charles B. Davenport and Albert G. Love, The Medical Department of the United States Army in the World War (Volume XV): Statistics, Part One: Army Anthropology *(Washington, D.C.: U.S. Government Printing Office, 1921), 41.*

Special Anthropometrics

Sitting height
Span
Height of sternal notch
Height of pubic arch
Neck circumference
Breadth of shoulder
Transverse diameter of the chest
Antero-posterior diameter of the chest, and thoracic index
Waist circumference
Arm length
Forearm length
Leg length
Knee height

Thigh circumference
Calf circumference
Suprapatella circumference
Knee patella circumference

Perhaps most illustrative for our purposes is the fifteenth volume, compiled by Charles B. Davenport and Albert G. Love, which focuses on statistics (and particularly part one, which examines the anthropometric data collected by the Army throughout the war).[94] The volume drew upon the anthropometric data recorded during Selective Service registration and mobilization and correlated it with medical data collected at hospitals throughout the war. These data were supplemented with additional anthropometric data that had been collected in 1919 during the demobilization of approximately one hundred thousand decommissioned military personnel (see Figures 2.15 and 2.16 and the list above). The new data included seventeen new measurements of things like thigh circumference and forearm length that were to be used to commission new uniforms, though officers in the Surgeon General's Office, in remembrance of the value of Dr. Gould's work during the Civil War, had been arguing for just this sort of detailed data collection since the start of the war, having only been convinced of the necessity of delay by the speed at which mobilization had to take place.[95] In comparing these measurements with Dr. Gould's from the Civil War, the statisticians were able to analyze mean changes in weight, height, and build, among many more minute measurements, for America's fighting men in both wars.

They also analyzed the data along racial and ethnic lines, looking, for example, at anthropometric differences between "white" and "Negro" troops. As Davenport and Love note:

> The general comparative picture we get of the white troops (including a great variety of races) and the Negro troops is this: The Negro troops have relatively longer legs and arms, shorter trunk, narrower pelvis, more nearly circular ellipse of cross-section of the chest; larger, shorter neck; more nearly parallel outlines of the trunk, larger leg girth, and a greater weight than the whites. The waist is less marked because of the relatively small transverse diameter of the pelvis and chest and the greater circumference of the waist. The

> Negro seems more powerfully developed from the pelvis down and the white more powerfully developed in the chest.[96]

The logical leap to eugenics-based arguments is obvious, and was quickly taken after the war as hordes of pseudo-statisticians utilized the anthropometric data and psychological data from the Army Alpha and Beta tests to further stabilize a discourse of scientific racism and ableism. These racial, ethnic, and anthropometric data were then further correlated to medical treatment data to look for correlations between stature, weight, chest circumference, and robustness with certain diseases or ailments throughout the war:[97]

The Dimensions of Men with Specific Defects and Diseases

Pulmonary Tuberculosis
Simple Goiter
Exophthalmic Goiter
Myopia
Hyperopia
Astigmatism
Hypertrophic Tonsillitis
Tachycardia, simple
Cardiac Hypertrophy
Mitral Insufficiency
Mitral Stenosis
Valvular Diseases of the Heart (Unclassified)
Varicose Veins and Varicocele
Hemorrhoids
Asthma
Defective and Deficient Teeth
Hernia
Enlarged Inguinal Rings
Flat-Foot
Defective Physical Development
Underweight
Overweight and Obesity
Cryptorchidism, Hypospadia, Anorchism, and Monorchism

The end result of these investigations was nearly two hundred pages of charts, graphs, and tables that examined correlations between anthropometric measurements, race and ethnicity, and instances of diseases, defects, and disabilities throughout the war. While it is nearly impossible to weave a narrative path through these reams of correlations across the data set, we would like to present some exemplary figures from the text and briefly outline their impact on future biopolitical forms of governmentality. In Figure 2.17 we can see a table presenting correlations between height and chest circumference and instances of tuberculosis. Tables like this were produced for all of the diseases, defects, and disabilities in the list above and their correlation with different anthropometrics. In Figures 2.18 and 2.19 we can see further aggregated data on height and weight distributions as they pertain to various special diseases or defects. These were also produced for each of the data points in the list above. Lastly, Figure 2.20 presents the highest-order statistical aggregation of the data, presenting instances of disease or defect relative to anthropometric measures where other measurements were held to the mean, such that the specific correlation of individual measurements to all instances of disease or defect might be analyzed in isolation.

This early-twentieth-century amalgamation of statistical correlations seems eerily similar to some of the claims regarding contemporary forms of "big data." Chris Anderson, editor in chief at *WIRED* magazine from

TABLE 140.—*Correlation between height and chest circumference (expiration) in recruits with tuberculosis (pulmonary), first (P_1) and second (P_2) million draft recruits.*

Height, in inches.	Total.	Chest, in inches. 28 and under.	29	30	31	32	33	34	35	36	37	38	39	40	41	42	43 and over.
58 and under	9		1		3	2	1		1		1						
59	12			3	4	3	2										
60	34	3	6	4	7	5	5	2	2								
61	64	2	7	11	17	11	11	3	1	1							
62	121	6	12	29	25	18	21	8	1					1			
63	236	2	32	41	50	41	33	18	13	4	1		1				
64	502	20	60	89	102	104	61	38	19	5	1	2	1				
65	782	14	74	158	179	151	107	57	24	11	4	1		1			1
66	1,192	22	92	210	230	223	212	111	44	20	6	1	1				
67	1,435	13	100	211	318	317	247	123	65	32	6	3					
68	1,579	14	99	206	330	328	274	171	94	41	14	3	5				1
69	1,489	17	64	176	275	351	285	178	89	30	14	7	2				
70	1,237	7	43	117	219	290	248	172	90	36	9	3	2		1		
71	884	6	26	70	122	200	185	141	80	42	11		1				
72	566		4	32	68	135	127	114	44	30	11	1					
73	282		5	15	30	62	96	54	28	27	3	1	1				
74	135		1	12	16	35	30	18	14	6	3						
75	53		2	3	8	11	6	14	4	2	3						
76	23				2	3	6	8	2	1	1						
77	5					2			2	1							
78	8			1	1		1	3		1	1						
79	1					1											
Total	10,649	126	628	1,388	2,026	2,293	1,918	1,233	617	290	89	22	14	2	1		2

P_1—
Number of cases: 4,627.
Height: Mean, 68.02 inches: standard deviation, 2.69±0.02 inches.
Chest circumference (expiration): Mean, 32.33 inches; standard deviation, 1.87±0.01 inches.
Correlation: 0.2391±0.0093.

P_2—
Number of cases: 6,022.
Height: Mean, 68.12 inches; standard deviation, 2.76±0.02 inches.
Chest circumference (expiration): Mean, 31.90 inches; standard deviation, 1.80±0.01 inches.
Correlation: 0.2499±0.0081.

P_1 and P_2—
Number of cases: 10,649.
Height: Mean, 68.07 inches; standard deviation, 2.73±0.01 inches.
Chest circumference (expiration): Mean, 32.09 inches; standard deviation, 1.85±0.01 inches.
Correlation: 0.2412±0.0062.

38636°—21——20

DIMENSIONS—TUBERCULOSIS.

Figure 2.17. Correlation between height and chest circumference with tuberculosis. Charles B. Davenport and Albert G. Love, The Medical Department of the United States Army in the World War (Volume XV): Statistics, Part One: Army Anthropology *(Washington, D.C.: U.S. Government Printing Office, 1921), 303.*

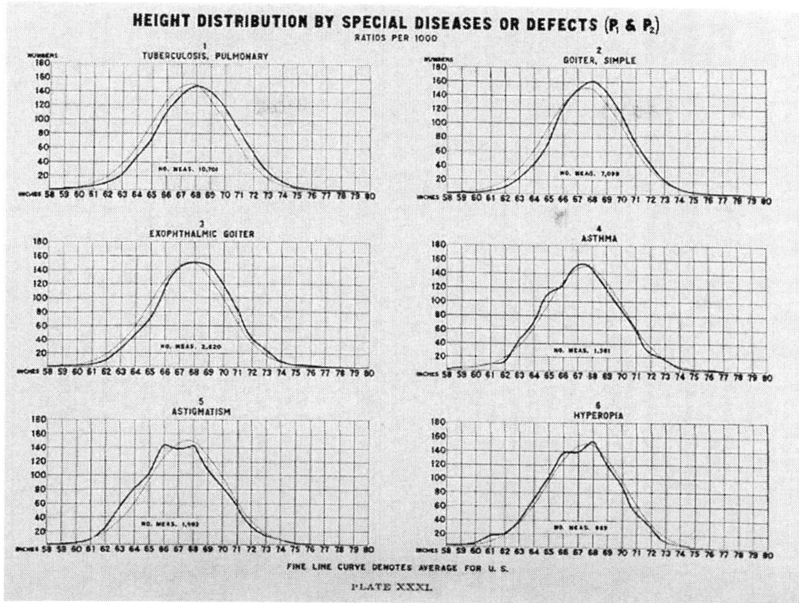

Figure 2.18. Height distribution by special diseases or defects. Charles B. Davenport and Albert G. Love, The Medical Department of the United States Army in the World War (Volume XV): Statistics, Part One: Army Anthropology *(Washington, D.C.: U.S. Government Printing Office, 1921), 386.*

2001 to 2012, famously argued in "The End of Theory" that science is no longer about rational humans using logic and empiricism to suss out causation, but is now about big data computation and the proliferation of non-causal, non-explanatory, but actionable correlations.[98] The future of epistemology is probabilistic and computationally rendered. It is ad hoc and restructures the games of truth that Foucault spent a career identifying. Knowledge becomes purely power, in the sense that its own truth is in its operational utility. We can already see this in embryo in the circuits we've examined from World War I. The Surgeon General's Office could not be sure what data would be useful or not, so it leveraged the country's momentum toward bureaucratization to capture as many data flows as possible throughout the war, preserving them all even when they could not be parsed quickly enough to be of immediate use. Even after war, there is no certainty about what data will be useful to future military endeavors. There isn't even any certainty about which data will have explanatory power for

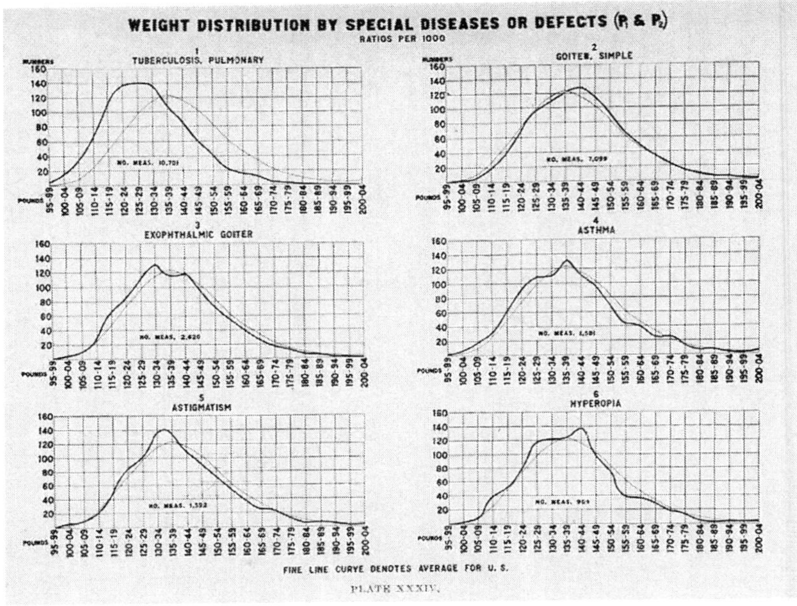

Figure 2.19. Weight distribution by special diseases or defects. Charles B. Davenport and Albert G. Love, The Medical Department of the United States Army in the World War (Volume XV): Statistics, Part One: Army Anthropology *(Washington, D.C.: U.S. Government Printing Office, 1921), 389.*

the past. So the answer instead becomes to render the data digital in its entirety, run it all through a computer—here still an assemblage of a tabulator and a (female) body—and publish all of the correlations it spits out. It is these probabilistic correlations that then become actionable "knowledge" for future efforts to produce the perfect soldier, mechanize decision making, and reduce human deficiency by eliminating defective military material and maximizing what remained.

The Impacts of the Selective Service Experiment

For Provost Marshal Crowder, the operative principle of Selective Service is what he terms "supervised decentralization."[99] Nearly seventy years before Gilles Deleuze would pen his "Postscript on Societies of Control,"[100] U.S. government and military officials were already envisioning such a future of networked command and control structures that could blend the type of

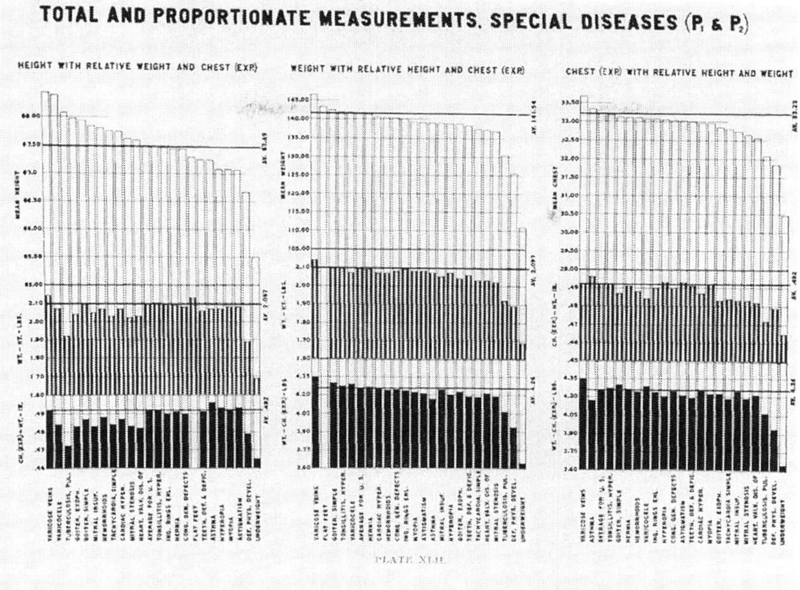

Figure 2.20. Total and proportionate measurements, special diseases. Charles B. Davenport and Albert G. Love, The Medical Department of the United States Army in the World War (Volume XV): Statistics, Part One: Army Anthropology *(Washington, D.C.: U.S. Government Printing Office, 1921), 397.*

dynamic local variability that creates a felt sense of self-determination with ongoing centralized omniscient governmentality. This networked form of governmentality was capable of eliciting maximized performances from the broadest spectrum of human potentialities at the lowest cost to the state while simultaneously allowing for the targeted elimination of unsuitable abnormalities, or "scum," according to Crowder:

> The Selective Service system was *continually* in process of change, and yet the idea underlying it never changed. It worked as the yeast works in the dough, or as the bacilli work in the cider, permeating the entire body politic and causing the scum to rise to the surface and be eliminated.[101]

We might better think of this networked form of governmentality as the power of the *circuit,* which utilizes dynamic addressability to continually modulate the circulation of bodies, supplies, techniques, and information. The production of massive, real-time, individuated or individually addressable data sets not only allowed for the statistical abstraction of governmental information for managing the population, but increasingly affixed a scarlet number to the subject's body, tethering it to a cache of individuated information ranging from archival data like performance metrics and medical and criminal records to statistical predictions like bodily capacities for education and labor and life expectancy. The temporal dimensions of the subject thus literally become an enclosed circuit etched according to the statistical modeling of governmental data. The subject's past, present, and future are written onto their encircuited body, or, what amounts to the same thing, their *files* in the cold, red ink of the state. Further, their indexical file provides the raw data upon which heterogeneous populations are produced, problematized, and governed.

The provost marshal general and the statistical department of the Surgeon General of the Army's Office rightly foresaw their work as the opening of a new problematic for governmentality and war. The past hundred years are rife with examples of this unique combination of supervised decentralization, big data, and triage, what we've here described as the power of the circuit. Much of President Franklin D. Roosevelt's New Deal policies and agencies were structured similarly, most notably the Social Security Administration, but also the Civilian Conservation Corps, the National Recovery Administration, and the Works Progress Administration, and mobilization for World War II utilized all of the data overviewed in this chapter as well as updated forms of the psychological assessments devised by the National Research Council in the draft. Similarly, we can see Robert McNamara's Department of Defense and the Vietnam War era projects like Project Igloo as logical extensions of the power of the circuit, as were President Lyndon B. Johnson's expansions of Social Security, most notably Medicare and Medicaid.

In the twenty-first century, the future of military triage most likely lies in the Trauma Pod, which researchers describe as "a rapidly deployable robotic system to perform critical acute stabilization and/or surgical procedures, autonomously or in a teleoperative mode, on wounded soldiers in the battlefield who might otherwise die before treatment in a combat hospital could be provided."[102] This research is being facilitated by the U.S. Army Telemedicine and Advanced Technology Research Center (TATRC), which is working with a number of private contractors and state universities to develop new technologies for combat casualty care, which includes "casualty extraction, physiologic real-time monitoring, and life saving interventions during the 'golden hour' while greatly reducing the risk to first responders."[103] TATRC oversees research into robots that can carry humans out of the battlefield, unmanned all-terrain vehicles and aircraft systems that can evacuate casualties, and advanced patient monitoring equipment like " 'smart stretchers' that allow for real-time physiologic monitoring throughout the combat casualty care process, from extraction to definitive care."[104] None of these are as sophisticated as the Trauma Pod, which companies like SRI International have been working to bring to the military since the 1980s. TATRC and its partners imagine the Trauma Pod as being capable of a wide range of functions, from telesurgery, and even automated surgery, to administering life support to advanced imaging and real-time machine vision analysis of results to shunts and laser cauterization to simple functions, like managing surgical tools, storing, depacking, and dispensing supplies (see Figure 2.21 for a list of Trauma Pod

SUBSYSTEM		**DEVELOPER**
SRS	Surgical robot subsystem	Intuitive Surgical, Inc.
AMS	Administrator and monitoring subsystem	SRI International
SNS	Scrub nurse subsystem	Oak Ridge National Laboratory
TAS	Tool autoloader subsystem	Oak Ridge National Laboratory
SDS	Supply dispenser subsystem	General Dynamics Robotic Systems
TRS	Tool rack subsystem	University of Washington
SCS	Supervisory controller subsystem	University of Texas
PRS	Patient registration subsystem	Integrated Medical Systems, Inc.
PIS	Patient imaging subsystem	GE Research
MVS	Machine vision subsystem	Robotic Surgical Tech, Inc.
RMS	Resource monitoring subsystem	University of Maryland
SIM	Simulator subsystem	University of Texas
UIS	User interface subsystem	SRI International

Figure 2.21. Trauma Pod subsystems. Pablo Garcia, "Telemedicine for the Battlefield: Present and Future Technologies," in Surgical Robotics: Systems Applications and Visions, *ed. Jacob Rosen, Blake Hannaford, and Richard M. Satava (New York: Springer, 2011), 40.*

subsystems in development). In short, "Fast and accurate triage, initial diagnosis of injuries, and stabilization procedures can all be accomplished remotely through advanced communication, telerobotic and robotic technologies. These technologies electronically project the specialized skills of the clinician to wherever there are wounded."[105]

What goes without saying is that these telecommunications are bidirectional, meaning that technologies like the Trauma Pod constitute new mechanisms for capturing real-time triage data from the field and monitoring an increasingly large number of physiologic data streams from casualties' bodies from the field to field hospitals to larger hospitals and medical institutions. The benefits are thus bidirectional as well, as soldiers gain the benefit of faster access to life-saving medical care while central administrators gain access to increasingly rich data streams that they can use to optimize the fighting capacity of the armed forces. It is not difficult to imagine a twenty-second century filled with personalized drugs and medical care driven by artificial intelligence (AI) that can be administered by miniaturized medical wearables or even internalized nanotechnologies. Imagine a soldier whose body and mind can be monitored in real time and modulated by twenty-four-hour automated injections of personalized cocktails of hormones, microflora, stem cells, and so forth to control for mood, alertness, neural plasticity, muscle mass, bone density, and so forth. Imagine triage administered by nanobots in real time or even in anticipation of predicted trauma.

Medical manipulation of the body, even by robotic AI pods, is still clumsy, slow, and too material when compared to remotely rewiring soldiers' biocircuitry as a means of maintenance and bodily revitalization. The Brain Research through Advancing Innovative Neurotechnologies (BRAIN) Initiative, partially under development by the Defense Advanced Research Projects Agency (DARPA) since 2013, is an ongoing attempt by the United States to more elaborately and immediately interface with the brain's circuitry. At least ten different subprograms are in active development for military purposes. ElectRx will use neuromodulation to rewire the already-existent biochemical circuitry in a way that is "automatically and continuously tuned to the needs of warfighters . . . reducing the time to treatment, logistical challenges, and potential off-target effects associated with traditional medical interventions."[106] ElectRx will further collect real-time data to provide a "quantitative framework to guide operations." This is merely one of many options on the operating table.

Here are a few more. The Hand Proprioception and Touch Interfaces

(HAPTIX) "program aims to create fully implantable, modular and reconfigurable neural-interface microsystems that communicate wirelessly with external modules."[107] Neural Engineering System Design (NESD) will build out "an implantable neural interface able to provide unprecedented signal resolution and data-transfer bandwidth between the brain and the digital world." Neuro Function, Activity, Structure and Technology (Neuro-FAST) will "enable unprecedented visualization and decoding of brain activity to better characterize and mitigate threats" while working to "facilitate development of brain-in-the-loop systems to accelerate and improve functional behaviors." Reliable Neural-Interface Technology (RE-NET) will "develop the technologies needed to reliably extract information from the nervous system, and to do so at a scale and rate necessary to control complex machines." Restoring Active Memory (RAM) will "facilitate the formation of new memories and retrieval of existing ones."

Where triage tents were used to perform life-elongating amputations, DARPA now supports programs successfully experimenting on mind-controlled bionics. "The prosthetic 'LUKE' arm system—which stands for 'Life Under Kinetic Evolution' but is also a passing reference to Luke Skywalker of Star Wars fame, who was endowed with a futuristic bionic arm—enables dexterous arm and hand movement through a simple, intuitive control system."[108] But the LUKE arm is merely a sub-experiment toward

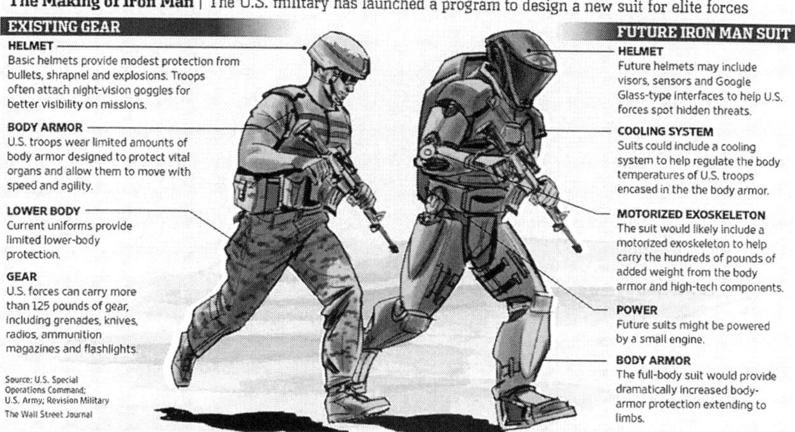

The Making of Iron Man | The U.S. military has launched a program to design a new suit for elite forces
EXISTING GEAR
HELMET
Basic helmets provide modest protection from bullets, shrapnel and explosions. Troops often attach night-vision goggles for better visibility on missions.
BODY ARMOR
U.S. troops wear limited amounts of body armor designed to protect vital organs and allow them to move with speed and agility.
LOWER BODY
Current uniforms provide limited lower-body protection.
GEAR
U.S. forces can carry more than 125 pounds of gear, including grenades, knives, radios, ammunition magazines and flashlights.
Source: U.S. Special Operations Command; U.S. Army; Revision Military
The Wall Street Journal
FUTURE IRON MAN SUIT
HELMET
Future helmets may include visors, sensors and Google Glass-type interfaces to help U.S. forces spot hidden threats.
COOLING SYSTEM
Suits could include a cooling system to help regulate the body temperatures of U.S. troops encased in the the body armor.
MOTORIZED EXOSKELETON
The suit would likely include a motorized exoskeleton to help carry the hundreds of pounds of added weight from the body armor and high-tech components.
POWER
Future suits might be powered by a small engine.
BODY ARMOR
The full-body suit would provide dramatically increased body-armor protection extending to limbs.

Figure 2.22. From existing gear to iron man suits. Defencetalk.com. U.S. Special Operations Command; U.S. Army; Revision Military; The Wall Street Journal.

more fully embodied bionic soldiering. Beginning in 1985, DARPA began work on an exoskeleton program. The interfacing envisioned by the various BRAIN Initiatives is in part a means of providing a perfect circuit that links soldiers and their bionic bodies.

While there is no single end point to the enlivened soldier in the circuit, we can make the quite obvious case that where sovereign modes of power threatened to extract the life of the soldier if he did not serve, and disciplinary power attempted to produce a level of docility that could overcome the fear of battle and the inhuman levels of suffering that take place in warfare, the soldier of the circuit will surely obey all orders as they can be programmed, modulated, given new memories, be spoken to by machines, psychoactively charged, and electrically stimulated to self-heal. The biopolitical circuit of security will erase any mode of resistance by simply plugging the soldier in to the circuitry directly through neurophysiological pathways that are capable of creating new memories. The body of the soldier is not surrounded and monitored by the circuit, as with surveillance; rather, biological processes are drawn into the circuit and are made to respond to its codes. The soldier becomes empowered in ways that make a World War I soldier lying in the trench with a single-bolt rifle seem like an archaic creature of another evolutionary epoch. Yet, the potential of soldiers to deny an order, go AWOL, mutiny, or even fill out a conscientious objector request will all be negated. If DARPA has its way, there will be no capacitor capable of short-circuiting the soldier.

3

Police Circuits

Render Automatic All the Mechanisms of Society

On June 30, 2018, in Villepinte, a northern suburb of Paris, five hundred grams of tri-cyclic acetone peroxide sat in the trunk of a car unexploded. The annual National Council of Resistance of Iran (NCRI) rally went as planned. An Austrian-based Iranian diplomat did not successfully mastermind an attack on the NCRI, a group that opposes the regime to which he pledges his allegiance.[1] In Tokyo, early 2019, a man suspiciously wandering through a retail store does not shoplift.[2] In Huntington Beach, California, during the 2015 U.S. Open of Surfing, teenagers did not drink alcohol or smoke weed in a parking lot adjacent to the event.[3] These non-events are policing in its purest form. Algorithmic policing based on a vast network of media used to collect data and sophisticated code to predict what "should" happen given the wrong set of circumstances ensured that nothing happened in any of these spaces at these times. The future of policing is already here and in this moment the *observer effects* are profound. For when your job is to police, and you can see the future, you have no choice but to change the world you hope never to observe.

While time travel might not yet be in the police arsenal, time manipulation has always been essential to the capacity of the police to control circulation in space. What follows, therefore, is a historical look at how the modern police force has been constituted through its technological capacities for time/space manipulation. From its birth pangs in the late eighteenth century, modern policing has been reliant on media technology to accomplish this task. As Peter K. Manning puts it, "The police face a communications problem: how to anticipate, respond to, mediate, and filter rising citizen demand with a controlled and calculated strategy. Police forces are based on complex communication systems and are dramatic, quasi-judicial organizations structured to allocate and deploy officers across time and space."[4] As such, police media encompass the technical and infrastructural means by which intelligence, personnel, resources, and patrol coordinates are collected, stored, processed, and transmitted through time and across space. Media provide a necessary condition for police departments and criminal scientists to produce knowledge: the surveillance and analysis of insecure spaces and human subjects can scarcely be undertaken without media that inform, constrain, and process that data. And not only are police interventions imagined and designed through existing technological capacities; they have historically been carried out by various logistical media, from the whistle[5] and call box[6] to the "wanted" poster[7] and police radio.[8]

For the purposes of this chapter, then, we will divide police media into two functional categories: police intelligence/surveillance and logistical telecommunications. The first addresses the role of media as mechanisms for producing databases, as police intelligence derives from the ability to amass, store, and process increasing amounts of data. These media work to solve the problem of time by extending the life of previously fleeting forensic information and opening it up to an ever-expansive field of criminology. The second function involves the extension, interactivity, and translatability of transmissions in order to initiate citizen surveillance and monitor and coordinate increasingly vast and mobile police forces. Such logistical media[9] solve the problem of organizing control over territory. Our analysis, then, will pivot upon what Peter Miller and Nikolas Rose have recognized as the essential interwoven procedures of governmentality: representing and intervening.[10] According to Miller and Rose, "The specificity of governmentality, as it has taken shape in 'the West' over the last two centuries, lies in this complex interweaving of procedures for representing and intervening. . . . We suggest that these attempts to instrumentalize government and make it operable also have a kind of 'technological' form. . . . If political rationalities render reality into the domain of thought, these 'technologies of government' seek to translate thought into the domain of reality, and to establish 'in the world of persons and things' spaces and devices for acting upon those entities of which they dream and scheme."[11] From this theoretical lens, the ways in which modern policing is constituted by technologies of government—that is, the coordinated use of media as technologies for intelligence gathering (representing) and logistics (intervening)—come sharply into view. Because the technologization of police activity has allowed the modern force to "police at a distance,"[12] police media have helped mediate between the police force's disciplinary tendencies and the liberal, circulation-biased political economies in which policing has taken form. In the words of Kelly Gates, this is a "balancing act that has consistently posed a challenge to liberal democracies and one that seems to lean, in the present climate, toward expanding police power."[13] As we will show, media have often been the tightrope on which this balancing act has been carried out.

At this time, critical/cultural scholarship addressing media technology and the police has largely been an affair of ideology critique featuring the analysis of police officers as characters in popular culture fare.[14] As this work has pointed out, popular police dramas have a long history that includes radio,[15] film,[16] and television,[17] and they have tended to support a "law and order" imperative. Similarly, most previous work on media and governmentality has fallen into two categories. First, a number of scholars have focused on the way governmental interventions are made via representational media content and the attendant guidance for spectatorship that teaches the audience how to view.[18] Second, several scholars have described the role of media technologies in the spatial and epistemological procedures essential to liberal government.[19] While the work of this latter group is sensitive to certain functions often associated with policing (surveillance and biometrics, e.g.), we would like to focus on how media have functioned in the modern police force: specifically, we are interested in how media are used to govern the police and police the population via the management of circulations.

While this is hardly an attempt to provide a comprehensive history of the police/media relationship, we do hope to provide several brief but illustrative glimpses into this history. In this historical overview, then, we will look at a small cluster of police rationalities and procedures from four crucial periods and trends in policing history. Our analysis begins with a look at how media like the police gazette were central to the fledgling liberal police project in the late eighteenth and early nineteenth centuries. Next we examine how anthropometric science and rogues' galleries provided two complementary media strategies for the mid-nineteenth-century police force. Third, we trace the relationship between automobility and the police, considering how the radar gun, the breathalyzer, the two-way radio, and other police media have been used to govern citizens' and patrols' automobility. Finally we delve into digital police media, which serve as a telos to the narrative of technological development we have been tracing throughout the chapter. The digital ideal—the ideal of the rapid and flawless storage, translation, and dissemination of evidence and other data—has long preoccupied the police imagination, and well-equipped urban police departments have often been on the cutting edge of technological developments. In conclusion, we argue that further analyses of the logistical aspects of police media allow for an understanding that may prove useful to contemporary political struggles, as the point of contact between police and resistant movements is largely organized and orchestrated by the use of police media in the streets.

Police/Media/Circulation

In the late twentieth century and beyond, Foucault argues, our notion of *police* has become clouded by the prejudices of the present. While we tend to think of the police as a specially equipped and legally privileged cadre of security personnel, the concept of *police* emerges from the liberal political experiments of eighteenth-century western Europe. Far afield from our current image of jackboots, semi-automatics, and billy clubs, the police arose from the liberal impulse to enforce a freedom of circulation "whereby everything would be controlled to the point of self-sustenance, without the need for intervention."[20] This ideal police state, therefore, would not necessarily be characterized by strong-arm repression or book burnings: it would, instead, "manage to penetrate, to stimulate, to regulate, and to render almost automatic all the mechanisms of society."[21] While this basic historical ideal has been overshadowed by the figure of contemporary professional law enforcement—with its crowning achievement in the birth of the *officer*[22]—early police theory's emphasis on securing circulation developed into the sine qua non of liberal social management. Police from such an idealized vantage were meant to be engineers and architects of the prison house of the circuit, not its prison guards. The following analysis, therefore, provides insight into how the police ideal and police practice collide.

For Foucault, in the eighteenth century the foremost police problem was ensuring "the spatial, juridical, administrative, and economic opening up of the town [and] resituating [it] in a space of circulation."[23] Early utopian projects for crafting the ideal territory deemed that the town, the capital, and the countryside should be architecturally arranged to ensure maximum possible circulation of people and goods. Police theorists, therefore, recognized that there was a spatial dimension to government. Thanks to technological and infrastructural developments in the late eighteenth and nineteenth centuries, the spatiotemporal domain of a given territory became manipulable in new ways. Accordingly, this manipulability fueled new strategies of governance that gradually displaced sovereignty's traditional methods of urban organization. Rail and electricity, for Foucault, established a new regime of circulation based upon the proper management of what he calls the "three great variables": territory, speed, and communication.[24] Managing these variables, according to Foucault, demands a strategy of governance less rigid than that of sovereignty, less confining than that of discipline—Foucault, of course, calls this emergent form of governance *security.*[25] While disciplinary power attempts to "isolate a space,"[26] to "prevent everything,"[27] and "allows nothing to escape,"[28] security involves "ever-wider circuits" in which "freedom is nothing else but the correlative of the deployment of apparatuses of security,"[29] organized through the "option of circulation."[30] What we see, then, is a practical ambivalence between an ideal disciplinary police power that seeks to control everything and an ideal police force that strives to efficiently manage flows and risks. In Foucault's words, it thrives by "allowing circulations to take place, of controlling them, sifting the good and the bad, ensuring that things are always in movement, constantly moving around, continually going from one point to another, but in such a way that the inherent dangers of this circulation are cancelled out. . . . It is simply a matter of maximizing the positive elements, for which one provides the best possible circulation, and of minimizing what is risky and inconvenient, like theft and disease, while knowing that they will never be completely suppressed."[31] The three great variables must be modulated and manipulated, especially by technical means, in order to organize, monitor, and regulate this circulation.

Rail, for example, brought new commerce and new goods, but it also brought new threatening subjects and populations. While in earlier centuries these threats might have been addressed by sealing borders or reinforcing walls, the principle of circulation lay at the foundation of liberalism's grand police experiment. A police society in this vein, therefore, would have to accept its share of vagrants, criminals, beggars, and the mentally handicapped; but it would, at the same time, develop new methods of surveillance, deterrence, and assistance to minimize the effects of these necessary evils. In short, throughout its history police society has relied on a system of switches and screens—on the development of technologies and procedures that not only facilitate movement but also foster the identification and measurement of threats, dangers, risks, and opportunities concomitant with that movement.[32] This sifting of dangers and risks endowed policing with a special mission within what Foucault calls "race war." For Foucault, this "war that is going on beneath order and peace, the war that undermines our society and divides it in a binary mode is, basically, a race war." This race war, according to Foucault, is based upon a recognition that "the social order is a war."[33] This detection of a permanent condition of warfare fortified the discursive, institutional, and eventually scientific construction of a "binary schema that divided society into two."[34] The "binary division" that characterizes race war provided the impetus for a society to organize against internal racialized elements. This division, which divided the population into an "us" and a "them," fueled political rebellions, slave revolts, and class warfare. But it also, of course, fueled the institutional development of policing, which was aimed at monitoring and containing dangerous classes and other racialized populations. And because—as we will see—policing in its infancy was as much a scientific and intellectual exercise as a practical one, its deployment in race war also led, as Foucault recognized, to "the production of fields of knowledge and of knowledge-contents"[35] that established and rationalized the differences between policed and non-policed populations.

The modern police force, perhaps more than any other public institution, demonstrates how liberal population management was characterized by these converging elements—modern media technology, circulation management, and scientific race war. Just as police societies ensure the proper spatiotemporal organization of goods like food and clothing, they also rely on media technology to seize the public's attention, measure deviance (and deviants), regulate traffic, and organize the bodies of its police officers. In the mid-eighteenth century, the classical school of penology put circulation and liberal population management at the center of its theories of crime prevention. Foremost among these theorists was Cesare Beccaria, whose Enlightenment ideals spawned a new penological science based upon reasoned, economical deterrence rather than brutal retribution. Beccaria plays a bit part in Foucault's *Discipline and Punish* as a penological reformer,[36] but his ideas on police and punishment go beyond Foucault's gloss. Whereas Foucault is interested in Beccaria's utilitarian theory of punishment, which instead of wreaking bodily retribution sought to discourage potential criminals into inaction, we are more interested in the policing implications of Beccaria's axiom that "It is better to prevent crimes than to punish them."[37] Describing an alternative economy of crime prevention, Beccaria writes: "It is not possible to reduce the turbulent activity of men to a geometric order devoid of irregularity and confusion. Just as the constant and very simple laws of nature do not prevent perturbations in the movements of the planets, so human laws cannot prevent disturbances and disorders among the infinite and very opposite forces of pleasure and pain. . . . Do you want to prevent crimes? Then see to it that enlightenment accompanies liberty."[38] By attributing the criminal impulse to human nature, Beccaria resigns society to a certain degree of criminality. It is better to accept the fact of crime, he writes, than to impose upon broad society an overtly disciplinary geometry of control that attempts to stifle human "nature": "What would we be reduced to," he asks, "if we were forbidden everything that might tempt us to crime?"[39] Thus we see an evolution in criminal response from retribution to surveillance and deterrence, such that the liberal policing apparatus is imagined in its capacity to prevent and detect crime while protecting citizens' "liberty." Beccaria's brand of enlightened liberalism, therefore, thrusts media into the center of the modern policing project, as they are used as a means to mediate between disciplinary control and liberal management.

Beccaria had a profound influence on the development of utilitarian thought, and he was especially influential on Jeremy Bentham's theories of policing.[40] Hence in the 1790s, when Bentham was commissioned to develop a policing method to prevent theft on the Thames River, he developed it upon the foundation of a Beccarian police economy based in preventive surveillance. With English police pioneer Patrick Colquhoun and Justice of the Peace John Harriot, Bentham devised a policing system that was unlike anything that existed in England, and indeed the rest of Europe, at that time. While amateur bands of watchmen armed with clubs and organized by horns and shouts had policed England's communities since the Middle Ages,[41] the idea of a sovereign, salaried patrol of police officers was unheard of; in 1798 there were fewer than one hundred police employees in all of England, and the majority of these were privately employed by West Indian merchants.[42] The advent of the public police force, then, was a pivotal and unprecedented development in European governance, and its exercise was at first severely constrained: in their early days, officers were even prevented from carrying weapons. The three principal factors that restricted the size and vocation of the early police force were, first, the rise of a liberal political order that transferred economic rationalities to the domain of governance; second, the fearful opposition that English citizens voiced to the rise of a sovereign police force;[43] and third, the profound demand that patrols made on public resources, forcing experts like Bentham and Colquhoun to devise ways in which media could manipulate the temporal constraints faced by a sparse, liberal police patrol.[44]

The rationality that undergirded the rise of the new police force was made clear in Colquhoun's treatise on one of the earliest policing experiments in England, the Thames River Police. Colquhoun used this 1795 work, *A Treatise on the Commerce and Police of the River Thames,* to explore how the increasingly restless working class could be better governed: "The mass of labourers became gradually contaminated. . . . The mind thus reconciled to the action, the offence screened by impunity, and apparently sanctioned by custom, the habits of pillage increased: others seduced by the force of example, and stimulated by motives of avarice, soon pursued the same course of Criminality, while the want of apposite Laws, and the means of carrying into execution those that existed, gave an extensive range to Delinquency. New Converts to the System of Iniquity were rapidly made."[45] Faced with the upheavals in politics and labor that occurred in the late eighteenth century, Colquhoun and Bentham devised an economy of police intervention that relied on logistical media and coordinated patrol efforts. Colquhoun writes that, in order to "renovate" the morals of the working class, the police force should be "aided by pecuniary energy, and by powers calculated, more to counteract the Designs of evil-disposed persons by embarrassing them at all points, than to punish. . . . And its effect will be the prevention of Depredations . . . in all situations where they were formerly committed. Upon this basis will of course be erected an improved System of Police Economy, in which will be combined everything that can tend to give utility and effect to the Design."[46] What was left to the police, then, was how to organize circulation in such a way to "embarrass" the designs of potential criminals.

Hence Colquhoun, Bentham, and other liberal utilitarians designed a policing system that reimagined criminality and crime response vis-à-vis flows of people and information. Police resources and personnel were distributed based upon the mobility demands of the fleeing criminal and the communication imperatives of a responsive/preventive police apparatus. Certain people, places, and activities were tied to what David Garland calls "criminogenic situations,"[47] which were generalized based on factors such as a location's capacities for communication and flow (lodge houses and horse coachmen, e.g.), locale (urban or rural), and the class and perceived moral quality of a business's clientele (bars were targets, e.g.). L. J. Hume describes how this process unfolded, showing that at the center of this liberal policing policy lay "an attempt to prevent offences against property by harassing receivers of stolen goods and by establishing a co-ordinated network of police authorities throughout the country. The attack on receivers consisted essentially in various measures designed to facilitate knowledge of the fact of an offence."[48] Thus one of the earliest rationalities of the liberal police force was that, in order to decrease criminal activity without a heavy police presence, two measures would have to be taken: first, the circulation of suspect and volatile populations would have to be surveilled and recorded; and second, certain lay individuals—such as innkeepers and carriage drivers—would have to carry out that surveillance by tracking and transmitting the behaviors of their fellow citizens.[49] Describing the advantages of implementing this vigilant lateral surveillance ethic among citizens, Colquhoun writes that the police should position itself against criminals in the same way a general plots for war: "Opportunities are watched, and intelligence procured, with a degree of vigilance similar to that which marks the conduct of a skilful General, eager to obtain an advantage over an enemy."[50]

In fact, some of Europe's earliest police officials were originally dispatched to regulate circulation via the maintenance of roads. In England in the mid-eighteenth century, towns received permission from Parliament to levy taxes in order to employ officials to pave, light, clean, and supervise streets. When "policing" is mentioned in official English documents from the period, the bundle of responsibilities attributed to the police officer always includes duties like "watchman" and "street keeper."[51] In 1795, Colquhoun, the founder of the Thames River Police—England's first official preventive police force—emphasized the importance of monitoring and managing the circulation of people: he proposed requiring all lodge houses to register their customers and enticing coachmen to become informants.[52] In a similar vein, at this time Bentham proposed a system of bookkeeping that should, he insisted, transcend the "pecuniary economy usually regarded as the sole object of bookkeeping."[53] Instead of keeping records for business purposes, Bentham saw in England's commercial system a promising site for carrying out surveillance on the population. Receipts and record keeping were no longer a means to record simple financial records; they were to be incorporated into a great police surveillance system. For Bentham, "Every significant transaction should be recorded."[54] Record books were thus reenvisioned as police media rather than means to document financial transactions, as fledgling police agencies appropriated traditional private media practices to trace the flows in and out of criminogenic spaces. Accordingly, in 1833, a mere four years after the founding of the modern police force in London, Parliament passed the Lighting and Watching Act, which allowed constables and their officers to maintain control over streetlights (an innovation forced upon constables two decades later with the Metropolis Management Act). In the face of rural England's slow adoption of the police force, Parliament convened a number of committees tasked with analyzing crime prevention. In 1839, a report on constabulary forces was penned in crisis because of thefts and banditry that had become increasingly associated with rural travel; the report hoped to discover "what kind of force would give confidence to travelers."[55]

Crowdsourcing and Circulation Management

In a related attempt to stifle the circulation of criminals and other undesirables, the earliest police bureaucracies used media to deputize local populations in the hunt for suspects. To foster this vigilance, police agencies would communicate rewards, crimes, stolen goods, and potential threats via newspapers, handbills, professional police gazettes, and strategically distributed "wanted" posters. For example, in the 1770s the "Hue and Cry,"[56] a police gazette devoted to the capture of criminals, deserters, and the mentally disabled, entered circulation; but it took several decades for this project to become successful throughout England.[57] In 1822, J. T. Barber Beaumont—one of England's chief law enforcement officers in the early nineteenth century—testified before a government committee about how to best overhaul the media-driven hue-and-cry system:

> It is true that there is a police newspaper called the "Hue and Cry," but it is only published ONCE in three weeks, and now that the communication all over the kingdom is so rapid, no one would think of giving three weeks start to a criminal before a hue and cry were raised. That paper is of no use. . . . To produce a really effective "Hue and Cry," it is therefore proposed for all informations

> of robberies, frauds, and other great offences in and about the metropolis, to be taken on oath at the police offices . . . [and] to be abstracted and transmitted every day at noon to a central office . . . [and] to be immediately inserted in a Police Gazette, published every afternoon and sent to every police office . . . to be by them filed.[58]

Beaumont relates that the central problems facing the early police are the spatiotemporal constraints that limit officers' abilities to intervene into and prevent criminal activity. The main problem with the "Hue and Cry" was its infrequency: with innovations in communication and transport, criminals were beating the police at the communication game. Old methods of police-media response were thus totally ineffective at meeting the challenges of the new century. A centralized communications office was needed, from which the news of all crimes could be transmitted to local jurisdictions each day. Further, these reports would need to be kept on file, so that offenses could be kept alive in the investigative imagination of local police bureaucracies.

This dream of perfect storage, access, and dissemination lends itself to demands for increasingly centralized police communications. When English politicians and bureaucrats were debating the need for a centralized police force in the early nineteenth century, committees were formed to assess the efficacy of centralizing England's sparse network of independent police jurisdictions. The 1816 testimony of Nathaniel Conant, the magistrate of the Bow Street Runners—a local policing apparatus that would soon provide the model for the rest of England—shows that centralized communications was a fantasy of the police even in its infancy. Responding to an examiner about the potential efficacy of a centralized office of police communications, Conant opines:

> I have seen myself that such a communication would have been desirable at one part of the town, of an offence committed at another; because they would possibly have discovered an important offender who was afterwards found to have escaped into that neighborhood. . . . I have often thought that such communication, but for the expense attending it, would have a beneficial effect. I have thought of recommending, that one of the Clerks of every Office, at three in the afternoon, should put upon paper a minute

> of important offences, and sent it in the Penny-post to each Office; it would reach the Offices in two hours, and would answer many useful purposes.[59]

Conant's testimony illustrates how in many ways the policing problem has historically been a media problem. The communication and storage of evidence and other data, as well as the deployment and coordination of patrols, have always shaped the aspirations and success of the modern police force. Because the constraints of mediated communication frustrate the policing project, the ideal of faster, more durable, and more accurate communication has proven to be a recurrent preoccupation of police officials working to maintain their traditional project of regulating and facilitating circulation.

Anthropometrics and Publicity in Nineteenth-Century Policing

As the police force continued to gain power and resources throughout the nineteenth century, its need for new knowledges and new instruments grew. To theorize the proper role of police in an economy of state practices, police councils turned to politicians and experts in the human and biological sciences, many of them practicing proto-eugenic specialties like anthropometry. As eugenics crept into most nooks of scientific inquiry, these experts and community leaders routinely reported that police could address growing crime rates—exacerbated by the industrial revolution and rapid urbanization—by linking citizens to vast surveillance databases that would contain photographs, family genealogies, and life histories. These databases were in part aimed at providing authorities with a vast resource with which they could isolate the genesis, habits, and physical characteristics of *Homo criminalis*, the (oftentimes constitutionally) criminal being.[60] More generally, anthropometric data were used to gather and store knowledge about how police resources could be best allocated in order to preserve the size and focus of the governmentalized police force. While the importance of anthropometric databases to eugenic criminology has been well documented,[61] we will turn our attention to how anthropometrics introduced new challenges and opportunities to the police media apparatus.

Media and modern forensic policing have long had a tight-knit relationship, and detective work and forensic science more generally benefit from the development of faster, more efficient, and more accurate methods for rendering residues of the past into analyzable data. In the 1880s, Alphonse Bertillon, who was one of the most influential police theorists in the late nineteenth century, introduced a new media-driven anthropometric science called "signaletics." Hoping to turn the human body into a technology of criminal evidence, Bertillon devised a complex anthropometric scheme that sought to identify criminals with empirical precision, and then keep representations of them on file and ready at hand. The rationality bolstering this development, writes one of signaletics' earliest American adopters, is to render the human body and its traces into data that could be harnessed by police agencies: "How much more precious still would such a means of identification be if it could be applied, not only to the living man, but to his dead body, even when crushed, mangled or dismembered beyond the recognition of his nearest friends and relatives!"[62] Although Bertillon's dream of a biometric passport system—which would require all citizens at all times to carry their "papers," complete with photograph, life history, bodily measurements, and fingerprints—never took strong hold in western Europe or the United States, his innovations in investigation, intelligence gathering, and surveillance have had a lasting impact on contemporary forensic and preventive policing techniques.[63]

One of the central problems that Bertillon's signaletics set out to solve was how to reliably record and circulate the identity of criminals. Although photography was an important part of his method, Bertillon recognized three shortcomings of a criminal identification system that relied primarily on photographs: first, Bertillon concluded that, given the state of photography at the time, photographs would have to be taken in the same place and by the same photographic equipment; second, he realized that photographs could not be easily reproduced and transmitted between jurisdictions; and third, he recognized that photographs captured superficial information that could be easily altered by clever suspects.[64] Signaletics, on the other hand, developed a thorough battery of bodily measurements that would quantify those parts of the adult body that remain relatively stable throughout life. Bertillon found that, if one were to take measurements of the head, feet, middle finger, forearm, height, ear, and other body parts, the statistical probability of false identification would be extremely low; and more important, these numerical anthropometric records would be much more comprehensive and transmissible than photographic images. So in addition to cameras and photographs, Bertillon turned to calipers, gauges, ink pens, and file cards in order to bring his "written portraits"[65] to life in a standardized form that could be widely and easily distributed.[66] Bertillon's signaletic science, then, was caught up in a broader media system environment that produced the constraints and constituted the possibilities of the policing project.

Signaletics soon gained immense popularity across the United States and in Europe and its colonies. In 1896, Boston's superintendent of police, Benjamin P. Eldridge, described his "mathematical" investigative strategy based upon Bertillon's signaletics:

> To facilitate the comparison and identification [of criminals], every set of measurements that is taken is recorded on a card, and this card is filed in a cabinet divided into compartments, each of which is subdivided. These separate compartments are used for the classification of cards in a way approximately resembling the filing of book cards in a public library. In searching to identify any person who has been arrested, the examiner . . . can turn at once to a compartment in the cabinet containing all the cards of persons whose heads come within this range. This compartment is divided into smaller compartments, each of which contains its special range of measurements of other parts of the body, or marks distinctions in the color of the eyes and hair. So if there is any card of measurements corresponding exactly to the set taken by the examiner, he will soon put his hand upon it.[67]

Eldridge illustrates how the late-nineteenth-century police strove to govern through setting out the "rascal" population as a manageable set of data. Yet this process of representation was severely complicated by problems of compatibility and speed; not only was the measurement of criminals and suspects a time-consuming and difficult affair—signaletics required eleven measurements—but scouring through archives for identity matches imposed a huge workload on police workers. As Eldridge explains, this was addressed by organizing a special filing system, as the desire for a more "mathematical" (i.e., digital) process preoccupied turn-of-the-century police strategy.

At about this time, an alternative criminal identification strategy, the rogues' gallery, rose in popularity in American policing. The rogues' gallery, which eschewed the private police files and note cards demanded by signaletics, gained prominence in the 1880s when Thomas Byrnes, head

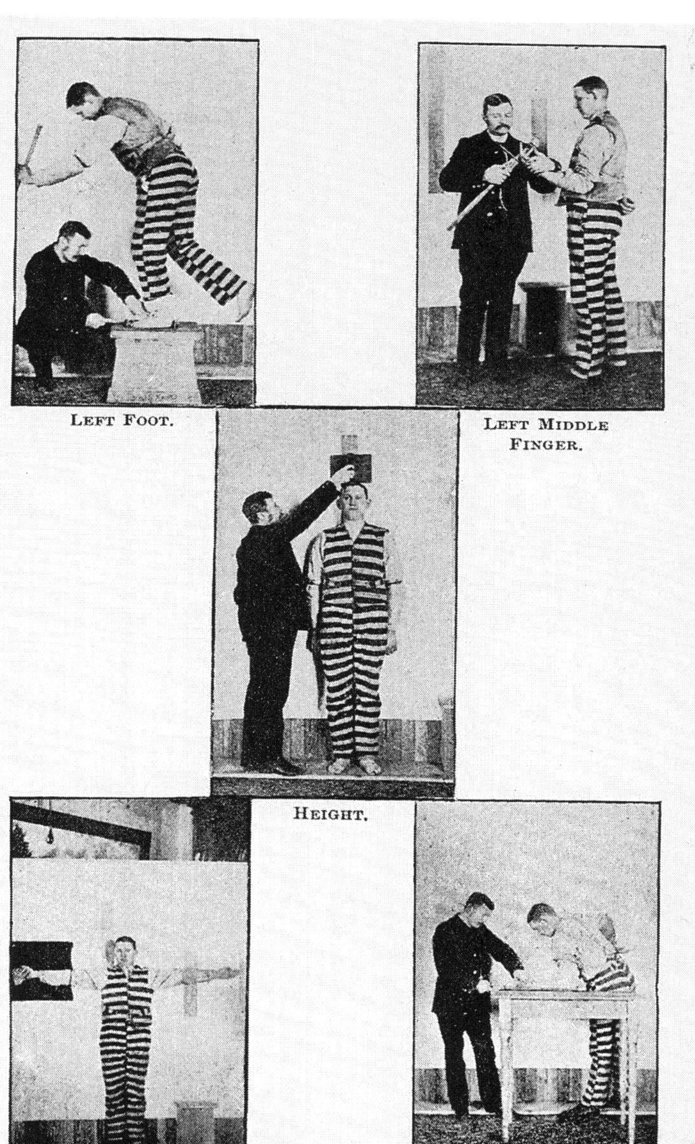

Figure 3.1. Demonstration of Bertillon's signaletic measurements. Benjamin P. Eldridge, Our Rival the Rascal: A Faithful Portrayal of the Conflict between the Criminals of This Age and the Defenders of Society—the Police *(Boston: Pemberton, 1896), 321.*

of the New York City Police Department, developed a criminal identification method that featured this gallery. Contra Francis Galton and other advocates of eugenics and the measurability of criminal "types," Byrnes cautions that theorizing a criminal class based upon physical features is a waste of time: "Look through the pictures in the Rogues' Gallery and see how many rascals you find there who resemble the best people in the country. Why, you can find some of them, I dare say, sufficiently like personal acquaintances to admit of mistaking one for the other. . . . In fact, it is a bad thing to judge by appearances, and it is not always safe to judge against them. Experience of men is always needed to place them right."[68] This recognition gave Byrnes a different set of concerns, as criminal identification was a matter with which nonspecialists could assist. With signaletics and similar file-based identification strategies in mind, Byrnes writes: "While the photographs of burglars, forgers, sneak thieves, and robbers of lesser degree are kept in police albums, many offenders are still able to operate successfully. But with their likenesses within reach of all, their vocation would soon become risky and unprofitable."[69] Byrnes brings to the fore the question of whether police should govern through public rather than confidential police data. For Byrnes, publicity had two advantages: first, it helped investigators and the public move beyond misleading stereotypes of the "criminal class"; and second, it effectively integrated representation and intervention into a single media process. Recognizing that rascals can appear as distinguished as any judge or businessman, Byrnes refused to use photography and anthropometrics to theorize the criminal body; instead, the rogues' galleries established the police's focus on the physical characteristics of the individual suspect. These photographs, then, did not seek to capture the physical evidence of a criminal's internal affliction, but were instead constitutive police media by which lay individuals were recruited into the policing apparatus. Like printed police gazettes and "wanted" posters, the rogues' galleries worked to deter crime by spontaneous recruitment and hence the threat of publicly distributed, ubiquitous surveillance. In this interesting case, message circulation doubled as recruitment.

We see, then, two distinct yet parallel ways in which police media were used to govern crime in the nineteenth century. First, the police strove to represent the criminal and his or her acts into actionable data, typically through the use of specialized personnel and investigations.[70] Today, of course, these media-driven practices still dominate the policing project,

Figure 3.2. Philadelphia's rogues' gallery in 1884. Howard O. Sprogle, The Philadelphia Police, Past and Present *(Philadelphia, 1887), 275.*

as technological developments such as fingerprinting, mug shots, and the polygraph shape police procedures at every level and allow the police to more efficiently circulate their virtual presence.[71] Second, police media were used to publicize suspects' identities, diffusing police responsibilities to the public and deterring crime through the insecurity of categorical suspicion and ubiquitous surveillance.[72] This method, as we pointed out in our earlier discussion of police gazettes, has a long history, and the value of this kind of publicity was hotly debated among nineteenth-century police officials.[73] Synthesizing practices of representation and intervention, this diffusion of policing responsibilities allowed the surveillance reach of the police to multiply while the force itself remained relatively sparse and inconspicuous. Taken together, these two different media strategies helped satisfy the creeping disciplinary ambitions of the police while permitting the liberal police force to govern best while governing least.

Media and Police Automobility

These principles of circulation ensured that patrol mobility and flexibility sat at the forefront of police concerns. The earliest forms of police patrol media were primarily mobile and included guns, lights, whistles, night sticks, bells, and other instruments that were used to draw attention, call to arms, and warn civilians of immediate threats.[74] Later, electronic systems connected the police to a broader network, but the media themselves were stable and tended to be placed in vicinities that were strategically and efficiently oriented according to logistical centrality and perceived necessity. In the 1850s, telegraphic call boxes were installed in some jurisdictions, allowing citizens one-way communications with the police.[75] More-efficient call boxes began to appear in the late 1870s in the United States, and by the end of the century they were distributed throughout the British Isles. From its inception in 1880 this hybrid medium combined the telegraph and telephone into a multiple-signal mechanism that could designate via coded messages a "fire, routine report, summon an ambulance, signal a riot, call a wagon or permit use of the telephone without a coded signal."[76] Of equal importance, "All signals are recorded on paper tape at headquarters, giving a permanent record of exact time and location of the call."[77] These new police media automated the collection of spatiotemporal data and provided a mechanism for surveilling officers themselves, an under-discussed yet essential element of the policing process.[78] Hence, as with so many police media, call boxes were logistical coordination and response devices as well as information-collecting technologies.[79]

This becomes especially important in the regulation of traffic and automobility. Consider, for example, how the goal of traffic policing is not really to catch rule breakers; it is to maintain efficient flows of people and goods. Infractions—speeding, for instance—are not a problem in and of themselves, but they do disrupt the system as a whole by creating a rupture that slows down, not speeds up, the average speed by creating excessive differences between vehicles. The perfectly policed highway would facilitate perfectly regulated circulation and would need no direct police presence. Yet in the absence of this perfect circulation, the police developed new means of enhancing their mobile capabilities on roadways. Transportation and communication technologies have played a central role in reorienting the police's relationship to the time/space axis.[80] Well before the patrol car was introduced, horse-drawn patrol wagons were widely established as a means of extending the range, speed, and load capacity of the police patrol. Although by the 1880s wagons were a mainstay of American police forces, their use restricted patrol to navigable roads, thus making the road a space of heightened policing.[81] Concerns with all three of these variables—speed, range, and load—continue to this day and are in part determined by the utilization of modes of transport as media technologies. Many early police cars were in fact an extension of the horse-drawn wagon, as they were envisioned as both roving criminal collection devices, whose enhanced storage capacity enabled the collection and processing of a greater number of criminals, as well as a means for the rapid deployment of a police swarm.[82] The San Francisco Police Department's official history focuses upon just such a transport/media "system": "In 1889 the department established a patrol wagon/call box system through which officers could call their stations for the first time and obtain speedy backup assistance. Reserve officers standing by in stations would mount the wagons and respond quickly to calls for assistance and other emergencies."[83]

These innovations in automobility gave the police a new set of opportunities, challenges, and preoccupations. In 1968, political scientist Paul Weston betrayed this sentiment, coining the neologism *motorthanasia* to describe the conditions that were leading to the United States' fifty thousand annual traffic fatalities: "The great tragedy of this age of motorthanasia is that the only forces waging a day-to-day fight against death and injury on the highways are the police."[84] Beyond instigating all manner of societal upheaval, the automobile radically reconfigured the nature of police work and soon became its primary security concern.[85] Widespread automobility altered the geometry of the "three great variables" by increasing the speed of transport for both criminals and police, extending the territory over which police patrols were distributed, and creating new demands and capacities for police communication. By 1932, with the advent of the first two-way radio system, mobile communication technologies further reoriented the time/space axis and helped create a new police media sensibility in which both too much communication and too little communication were considered a threat to automotive safety.[86] As has been shown, the question of controlling dangerous mobilities had been a concern of police since its infancy, so these developments are more a change in scope than one of kind. Further, while automobile-based police patrol had become a mainstay of U.S. policing by the 1910s, it was by no means the only form of patrol, nor was technologically enhanced mobility an entirely new element of policing. Foot, horse, bicycle, motorcycle, boat, and aircraft were all prominent modes of police transport by the 1960s,[87] thus making the police's territorial reach and intelligence capacities both extensive and intensive.

This extension of mobility provided new means of gathering intelligence and enacting surveillance. For instance, radar guns were introduced in the 1950s to record the speed of vehicles, allowing for (1) individual citations to be given according to strict, mechanically collected data free from the subjective assessment of police judgment, and (2) the generation of a broader database of speeding at the level of the driving population. In this rudimentary sense. radar detectors were both a technology of representation and intervention. In addition, airplanes eventually allowed for the ability to monitor a greater number of speeders while simultaneously providing a "bird's-eye view" of the motoring population, seen as a mass or aggregated traffic system. Thus police airplanes are both modes of transport and "ways of seeing," as cockpit windows create a unique screen through which to monitor the world. Hence, at the level of police intelligence these newly integrated technologies extend the range of surveillance and offer new forms of data for circulation, collection, storage, and analysis.

To combat the most vilified form of illegal automobility, drunk driving,

Figure 3.3. Police patrol wagon, 1880s. Baltimore City Police Department.

Figure 3.4. Police patrol wagon, 1920s. Detroit Public Library Digital Collection.

a number of media were brought into play in order to validate officers' suspicions. In the 1920s, Indiana University's Rollo N. Harger developed blood and urine tests that police could use to measure the alcohol content of suspects' bodily fluids; but because these tests took several days to process, Harger and his team developed new ways to increase the speed of alcohol detection.[88] By the early 1930s Harger developed a set of subjective criteria by which specially trained officers could detect drunkenness in suspects; yet while this met the demands of speed, it left much to be desired in accuracy. Then in 1938, Harger invented what he called "the Drunkometer," which had its own problems: it was clunky and had to be recalibrated each time it was moved, so while it was a useful medium for measuring intoxication levels in laboratory studies, it was of little use to a mobilized police force. Yet in 1954, another Indiana University professor invented the Breathalyzer,[89] which has become the police standard for collecting intoxication data, providing stable results even while being transported in an automobile. These police media function by turning phenomena into measurable illegalities. Risk was thereby made indexical, and contingencies are given the veneer of scientific determinacy via quantification: "drunkenness" suddenly becomes a quotient, 0.08 blood alcohol content. In other words, the demand for objective criteria in the assessment of risk demands media for turning worldly phenomena into measurable and hence governable data.

In 1968, prominent criminologist Paul B. Weston wrote that "a successful tactical plan for action by police and other agencies concerned with traffic safety can only be based upon the facts confirmed in these [traffic] records. . . . [Yet] there has to be a cutback in the mass of accident records maintained by police . . . so that accident records can be maintained in some manageable form."[90] As with Bentham and Colquhoun's approach to policing the Thames, traffic management would 150 years later follow the same sets of concerns, such as "identify dangerous locations or areas, . . . time of accidents, . . . cause of accidents, . . . [and] reveal the effectiveness of police traffic control activities."[91] Ledgers, carbon-copy accident report forms, specialized training for accident assessment, weekly and monthly trend reports, location-based filing, spot maps for high-frequency accident locations, lists of most hazardous locations, collision diagrams, specially outfitted accident investigation cars, and "an efficient accident records unit"[92] have all been deemed necessary elements of a successful police media apparatus.

After the Unsafe at Any Speed movement,[93] governmental response to traffic safety invested an unprecedented amount of resources into studying traffic safety and implementing direct and cybernetic interventions. In particular, a variety of police media were subjected to assessment and quality control, including rumble strips, school zone flashers, freeway television surveillance, traffic control by radio communication, nuclear-energized self-luminous highway signs, in-car driver warning devices, computerized traffic-actuated signals, communication of disabled vehicles, electronic traffic control and surveillance, and computerized traffic simulation modeling.[94] Treating these technologies as police media allows us to understand the nuanced ways that communication processes are elaborated and automated as a means of altering conduct in an efficient and unintrusive fashion. Consider the rumble strip, which is a cybernetic police agent that responds to a driver's communicative action—for example, "I'm not paying attention"—with its own response, "Pay attention! Steer back on the road!"[95]

While driver's licenses and automobile license plates function as remote indicators of identity, these technologies provide little data in and of themselves; they are fully operative only when they become indices within a broader data set of the criminal record. Further, the remote retrieval and input of data while on patrol spreads the effectiveness of police media as collectors and processors of information; the network is thus enlarged and sped up due to mobile media, and net power is engorged. It is no surprise, then, that the police have been at the forefront of mobile media. While militaries have generally outpaced police in their use of media at the tactical level of managing space and quickly amassing force,[96] police have led the way in mobilizing remote databases. Police media have been making these data remotely available for hundreds of years through various print media, and with the advent of the telegraphic call box access became markedly smoother and more immediate by the 1860s. However, more remote forms of radio, telephonic, satellite, and ultimately digital media have steadily increased the spatial and temporal frontiers of the modern police force.

Police and the Digital Ideal

Even though the broad-based amassing of data has some imagining Big Brother scenarios of all kinds, police administrators still use the language of efficiency and intelligent application of limited resources. The police record is not supposed to be all-encompassing, but rather intelligently shared by necessary policing agencies through translatable and accessible channels. Concerns over data loss and incommensurable media systems plagued police telecommunications specialists for decades. How accurate was a 1924 fingerprint sent from New York to Chicago by Western Union's Telepix system? Analog records always had to be translated through transmission. This could mean a police communications officer accessing a criminal file in a cabinet and then reading important data to officers in the field listening on their one-way police radio. Or it could mean sending a "missing persons" photo through a telephone-assisted facsimile apparatus, a much more complicated multi-sequenced translation process that begins with an arbitrary flashbulb and an emulsifying chemical bath; decades later that result could travel through the phone lines as binarized data to be reconfigured as a pixilated image to connect a corpse with a criminal record continents away.

Incompatible systems and inaccessible databases were the two main problems confounding police omniscience, as data were in the wrong medium or were not properly mobile. To a large degree, such media incompatibilities were blamed for the failure of police to stop the 9/11 attacks.[97] More broadly the Department of Justice's (DOJ) IT plan presents a world of complexity laden with potential security threats:

> Terrorist attacks, natural disasters, and large-scale criminal incidents too often serve as case studies that reveal weaknesses in our nation's information sharing capabilities. Current information collection and dissemination practices have not been planned as part of a unified national strategy. A tremendous quantity of information that should be shared is still not effectively shared and utilized among communities of interest (COIs).[98]

Such DOJ documents are filled with infrastructural plans, diagrams, and flowcharts that clarify the necessity for a networked system that is still oriented around a central force, the U.S. government. As they state, "The Department supports both centralized and distributed models for information sharing."[99] While such a system has many nodes and is distributed, it is not a uniformly equal system, but corresponds to the asymmetrical power relations natural to most networks.[100]

As this general state of impending security crisis has become the "new normal," many calls have been made for fully networked and entirely digitized police media at the regional and local level as well. The DOJ calls for just such media in its aptly titled policy brief "Effective Police Communications Systems Require New 'Governance.'"[101] The brief describes the history of police media incompatibility as originating with the use of police radio in 1933. Various interregional and extraregional law enforcement and public safety agencies created separate and incompatible communications systems; hence these "stovepipes" were left both ineffective and inefficient. Instead the DOJ calls for complete interoperability in which agencies share ownership, control, access, data, and costs. Similarly, by 2010 at least seventy-two "fusion centers" had been created by the Department of Homeland Security in order to facilitate the collecting, sharing, and processing of data by local police forces, state police agencies, the FBI, and other intelligence services.[102] These centers point toward trends to more broadly network police media with all digital media. Digital information is gathered from all possible points at these centers, in part by private industry data specialists, in order to profile potential criminals, terrorists, or political dissidents.[103] Social-media posts and videos are treated as criminological data. Such policing of media could exponentially broaden the consideration of what may count as police media. It is at this point that we see the tendency toward "surveillance creep"[104] that in this instance aligns with the more disciplinary or all-encompassing form of policing. Fusion centers are light, efficient, and relatively unobtrusive, described as being "soft surveillance,"[105] thus looking an awful lot like the kind of "governing at a distance" most associated with modes of governmentality and security.

The vast scope and intensive data processing of these practices raise considerable questions when one considers that all data are potentially criminological data. The ideal of digitality is to make all media one medium.[106] Further, with the advent of internet-enabled mobile media, perfect and seamlessly shared police knowledge can theoretically be collectively captured and processed everywhere, anytime, and by any police actants. The world can be turned into digitized data through numerous police media, such as radar guns, Breathalyzers, CCTV cameras, digital fingerprinting, as

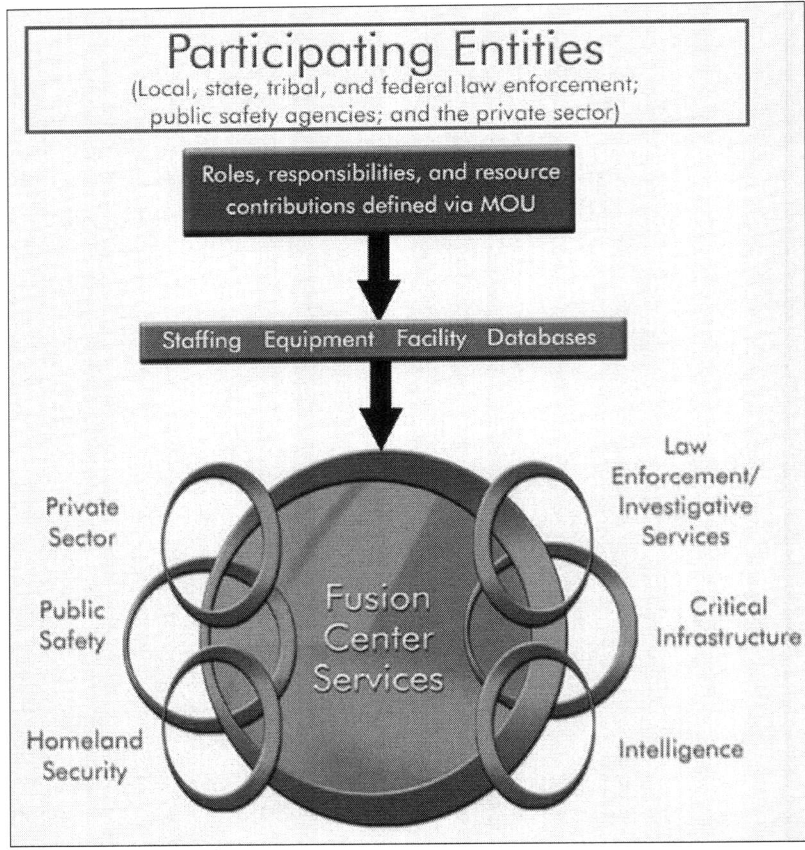

Figure 3.5. Department of Homeland Security partners. NSA.gov.

well as by "deputizing" all digital media in vast data-veillance efforts. Biometric, photographic, economic, demographic, genealogical, geospatial, and unspecified digital traces all exist as algorithmic and probabilistic potentiality. Temporal and spatial dynamics can be accessed, assessed, or even randomly generated to model optimal policing procedures.[107] The logic of security, unlike disciplinarity, is not binary—separating out the desirable from the undesirable—but it has increasingly come to depend on turning the world into binary data.

Conclusion

We want to conclude by offering two suggestions, one scholarly and the other political. First, there is much to be gained through broadening the analysis of media to include logistical media, especially as it concerns governmentality. The technical and logistical aspects of governance are very often served, promoted, and enabled by such media. Our analysis of police media is merely a thin slice of one arena of how logistical media make modern governance possible. Importantly, the considerations of representing and intervening retain their importance for both historical and present-minded critical research. Redirecting analysis toward how media function in the circulation of bodies and the production of knowledge, as opposed to the dissemination of knowledge or untruths, seems like a good starting point for broadening our understanding of how media function within different modalities of governmentality.

Second, it is important to recognize that police resistance needs to attend to police media not simply in terms of ideological maintenance but rather as strategic and logistical mechanisms for the collection, maintenance, and distribution of information and force. The police are increasingly looking to monopolize media by making illegal the tactical and logistical use of media by resistant forces. As Jack Bratich has explained, the ability of radical protest movements to deploy media tactically has often been prevented through incarceration or media incapacitation, as we saw with some Black Lives Matter protests in 2020.[108] In such instances the police maintain superiority not simply through a monopoly on the use of violence but by creating a monopoly on the use of logistical media as well. While a populist ideology of the digital heralds social media as tools for the organization and radicalization of resistant bodies, these same tools are just as easily used by the police for identifying protestors, infiltrating dissident groups, and coordinating counter-tactics. This police co-option of media technology even extends to devices that are meant to check police power, such as police body cameras. Although many activists have called for the introduction of these body cams, relying on media worn and controlled by the police also presents new challenges. While many view these cameras as mechanisms of police accountability, they can have the opposite effect by providing a veneer of transparency for cops who find ways to hide their brutality and other illegal conduct. In many cases, body cameras push crooked cops into a "complicated dance"[109] whereby they develop well-honed strategies for eschewing this media-assured accountability—for example, by retreating into spaces of extreme unintelligibility, such as interrogation rooms or police station bathrooms, to assault arrestees. Such tactics have also come under increased scrutiny due to privacy concerns and the inequitable targeting of nonwhite populations and individuals. Many tech companies, such as Amazon, Microsoft, and IBM, that had previously been providing AI-enabled facial recognition technologies to police forces quit doing so following the Black Lives Matter protests of 2020.

Despite these apparent rollbacks in AI-enhanced police tactics, the terrain of policing itself has become increasingly digital. International and domestic cybercrime has replaced burglaries, traffic infractions, and vandalism as a predominant police concern. As such, AI isn't only used in the application of "real-world" policing; rather, AI and other forms of problematic digital circulation are now themselves the terrain that must be policed. Here we see the logic of policing as the logic of circulation having to remake itself in a fashion similar to its remaking at the outset of the age of automobility. It is not only that digital media can be used to aid in committing or fighting crime, but rather that digital (cyber) practices themselves are criminalized. If we think of the space of criminality as bounded by the territorial maps that mark out police jurisdictions (city police, state or provincial police, federal officers of the FBI, FDA, RCMP, Scotland Yard), cybercrimes produce vastly different circulatory threats and force a reimagining of jurisdictional boundaries. There has of course been an ongoing struggle over the ways in which the media tools of policing have themselves been criminalized if used by non-police actors (radar detectors, police scanners, citizen video surveillance, sousveillance, etc.). In this regard, we should keep in mind that well-meaning calls to "defund the police" will almost certainly result in the replacement of cops with police robots, predictive policing, and more pervasive and invasive surveillance technology. Needless to say, it is unlikely that these labor-saving "police media" will solve the policing problem.

Accordingly, physical violence via police media should be a central consideration for protest movements. As Greg Elmer and Andy Opel clarify, such tactics transfer to the homeland the preemptive logics that spawned the Iraq War, as media are engaged domestically against dissidents to stop them before they even begin to protest.[110] Weaponry such as tasers are ideally suited for what are always-already-mediated events such as protests, because they leave no visible trace of the violence they perpetrate and on video footage look relatively harmless even as they are being deployed. Further, some police media are violent in and of themselves and also fail to leave a trace. The 2010 G20 Summit in Pittsburgh saw the inauguration of sound cannons, which fire a sonic blast of up to 150 decibels, producing intense pain for those in its path—what Patrick Gillham, Bob Edwards, and John Noakes call "strategic incapacitation."[111] After their success against G20 protesters, these cannons became a routine part of the police-media arsenal: they began to be deployed extensively in the early 2000s, and they have been used during the 2020 racial justice protests and subsequent scraps between cops and activists. Police agencies seeking to disperse threats have increasingly turned to LRADs—long-range acoustic devices—which are military-designed weapons that "fire" incapacitating sounds more precisely than traditional speakers.[112] In the 2020s, LRADs and other "critical crowd communications" systems have become a pivotal site of tactical police/military crossover as America's wars turn increasingly domestic.

Understanding police media, then, is not simply of interest to media historians; rather, it is necessary for understanding how and under what conditions media are policed—how they are made part of the policing process and how their use by protestors, criminals, and other potential enemies is suppressed, appropriated, or made illegal. As digital media become increasingly central to policing the public (and, of course, to policing dissent), the logistical mechanisms by which policing takes place must not only be made transparent but need to be understood in order that countermeasures can be devised.

The age of the body cam came and went. In the end it was deemed a failed media solution trying to account for a media resistance problem. In an age of precognition, the body cam was anachronistically post facto. It served to produce evidence for the realm of juridical punishment but failed to birth the "eye of power" central to Bentham's panoptic fever dream. Though it was often forgotten, central to Bentham's vision of surveillance as the most perfect form of power was the fact that the watchers, the guardians of the prison, were also placed under perfect surveillance. Their every move could be monitored and honed to produce ideal guardians. Reforming the guard–prisoner relationship or the police–criminal couplet meant not only reforming and reworking the criminal but reforming the police. The perfect police state would necessarily be filled with perfect police officers: the perfect prison, perfect guards. The body cam was a far-from-perfect technology.

In the age of the prison house of the circuit, perfect circuitry could theoretically produce predictive policing, but could it produce perfect police? Would the circuitry work to weed out the racists, the egomaniacs, the bullies, the MAGAs, and the jackboots from the force? Or could a chair of torture be built to reformat and rewire the police brain, neuter officers' capacity for throat stomping, misidentifying wallets for weapons, and seeing a threat in every dark face? Might predictive policing be best served to police the future of policing? These were questions tasked to the Predictive Investigations Group Pioneering Engineered Neutering (PIGPEN).

PIGPEN realized that a prison house of circuitry needed to encircuit the police as well. PIGPEN could realistically turn its predictive capacities upon the police force to determine when and where and which officers were likely to commit criminal acts, but the fundamental question around proper policing as such could not be answered via the logics of engineering and efficiency. When police brutality was central to the maintenance of racial and class-based inequity, what police behavior couldn't be deemed criminal? When police files and "intelligence" existed to ensure there that was always adequate evidence to maintain crippling incarceration rates for Black, Latino, and poor rural white populations, then which police media weren't creating false positives? When the law of the land was based on contractual obligations to property, then what ultimately were the police serving and protecting? Such problematizations paralyzed PIGPEN's capacity to engineer a perfect police officer. No police media were adequate to this task. No policing of the police could ever fix policing.

4

Circuitous Maximus

Automobility, Flow, and Driverless Futures

Rome's Circus Maximus was built more than twenty-five hundred years ago to host up to 150,000 spectators who watched chariot racing, gladiator battles, and numerous other religious and civic rituals that structure the cyclical rhythms of civic temporality. Chariot racing shared many features of modern circuit racing: mass spectatorship, an oblong track, barriers to shield spectators from the elements, and gruesome accidents. It was a place where people went to satisfy the need to be entertained, to pass the time, to watch things go fast, to witness the limit point beyond which controlled centrifugal velocity careened into death. As part of the second half of "bread and circuses," it performed an important role in keeping the populace in a social, religious, and political circuit.

Presently, vehicular racing takes place in a host of terrestrial, aquatic, aerial, and digital circuits. While NASCAR and Formula 1 may seem to suck up most of the North American oxygen, the Canadian Snowcross Racing Association provides a circuit for snowmobiles, Super Boat International hosts watercraft races, the Federation Aeronatique Internationale oversees competitions for all manner of airborne vehicles from vintage fighter jets to helicopters to drones, while Roborace dispenses with drivers as a mechanism to develop AI-guided race cars. As a plethora of vehicles trace the bound circuit, they remain closely monitored. Their feats are recorded, processed, analyzed, and circulated by broadcast, stream, and other means. Between the television broadcast and television viewer is another circuit that brings the race out of the space (and sometimes time) of the track to any number of speakers or screens. The larger circuit is one in which mobility, economy, and entertainment function as they did in ancient Rome in a broader celebration of technology, risk, and collective participation: blowing off a little steam.[1] But this relation isn't purely celebratory. The Circus itself was an exemplar of the logic it exhibited, with

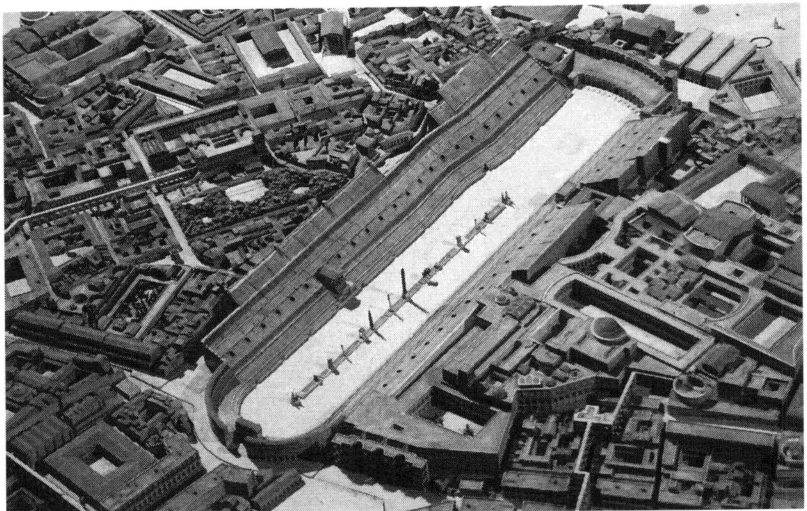

Figure 4.1. Circus Maximus. Created by Pascal Radigue / CC BY-SA 4.0 (via Wikimedia Commons).

"entrances and ascents for the spectators at every shop, so that the countless thousands of people may enter and depart without inconvenience."[2] Much more profoundly, Rome's political power was intricately tied to the circuits of roads and bridges with which it had crisscrossed its empire. The feats of speed and military prowess on display at the Circus bore relation to the speed and security at which currency, troops, governmental administrators, and encrypted directives circulated throughout the empire. The Circus was spectacle and laboratory, a theater for experimenting with technical prowess and populace alike.

While policing and automobility have been intricately linked for over a century, as examined in the previous chapter, the cultural, economic, and technological orientations of automobility provide a very different lens for examining contemporary circuits as they organize daily life and lifelong social dynamics. Specifically, an analysis guided by two interrelated concepts, *mobile privatization* and *flow,* first developed by Raymond Williams in his groundbreaking 1974 book *Television,*[3] provides a media genealogical approach for understanding how automobility and mobile media have aligned as mechanisms by which circuits work in concert with one another.[4] To be more explicit, Williams's concepts of flow and mobile privatization provide valuable insights into how circuits are interconnected and how they organize and manage spatial and temporal orders. In *Television,* broadcasting serves as a form that makes the cultural technology "television" possible, which in turn serves as a junction through which power flows and the viewer/citizen/consumer is organized. This organizing is accomplished in and through television in much the same way that mass media such as print, film, and radio broadcasting had done in the past, with the advantage of combining audiovisual engagement, simultaneity, and perception of choice. In the post–World War II boom economy in the United States, automobiles, home appliances and electronics, and a flow of audio and audiovisual programming over the air served as a virtual plaza where citizen–consumers could be organized and understood en masse. More recently, gains in transmission, speed, bandwidth, and processing power have enabled ever-finer-grained management of these circulations, to the extent that individual users can be more clearly monitored, surveilled, profiled, and advertised to at any point in space and time. Today's flow is less like a river or a spigot and more like the flow of electrons from one orbit to another: you can see where the flows are but not how they are moving; you can see how they are moving but not where they are. Media escalation has made life more connected, efficient, and continuous as circuit logics increasingly perforate time and space, enabling users nearly universal access while simultaneously rendering them accessible whether they are engaging or not. *We are the flow in the circuit.*

This chapter addresses the rise of autonomous automobiles and the increased means by which media come to be a mechanism for organizing the spatial and temporal dimensions of mobility in North America. We suggest that the period that Williams described in the early 1970s—the era of broadcast media, which runs roughly from the 1920s to the 1990s—exemplifies a disciplinary logic for the integration of mobility into the circuits of economic production and consumption for industrial–consumer capital, as well as the social norms associated with the postwar period. Beginning roughly with the introductions of CD players and car phones in the late 1980s and early 1990s, we begin to see a new orientation toward networked and digital media forms proliferating in the automobile, and these media become the backbone for imagining the kind of autonomous automobile system promised by automotive manufacturers like Tesla, tech giants like Google, and online retailers like Amazon. These autonomous systems are idealized versions of how flows operate in the age of information/creative/postindustrial capitalism, which has unhinged the rigidity of disciplinarity's spatiotemporal demands and potentialities by reorganizing circuits as elaborate mechanisms of flexible circulation. The following historical investigation demonstrates how the automobile has long been a testbed for the governance of circulation and encircuiting, where embodied movement that was once its central tenet appears to first be managed by media forms and then potentially replaced altogether as some imaginative futures envision.

Here we refocus our efforts to address *mobile privatization* in light of recent developments, from ride-sharing services that condition drivers to become passengers to emerging economies of driverless vehicle platforms. We take a specifically media genealogical approach and retain Williams as a critical voice. Although he did not set out to do media genealogy, it is difficult to argue that *Television* is not historical, critical, immanent, experimental, and problematizing.[5] Williams undertakes his project with care, warning against historicizing the relation between technology and society that often results in a series of deterministic perspectives.[6] Examining the telemobile circuit in the present demands an updated and materialist understanding of flow that attends to how the logic of the circuit creates free subjects amenable to neoliberal configurations of governance and capital.

Williams describes the technological changes that took place in the mid-twentieth century as a form of *mobile privatization,* "an operative relationship between a new kind of expanded, mobile and complex society and the development of a modern communications technology."[7] But what is a *form* of mobile privatization? What kind of a thing is this, and what can we do with it in a media genealogical framework? We suggest that mobile privatization can be productively deployed as a problematization, a dispositif, a discourse network, a background against which present automobile, televisual, and mobile communicative practices can be more productively understood. According to Williams, communications technologies are first and foremost developed to solve "problems of communication and control in expanded military and commercial operations."[8] Previous chapters have focused on communication and control in warfare and in policing, and that focus shifts in this chapter to sites of personal communication and commercial consumption. Specifically, we position screening technologies as strategies the mobile subject and various institutions deploy to negotiate communication and movement in a world made increasingly crowded, fast, and noisy.

Our focus in this chapter is less about specific existing, proposed, and imagined systems designed to manage mobility and more about the conditions of possibility for this management through flow and the increased potentiality for mobile privatization. In other words, how is governance reimagined in accordance with new technical possibilities (imagined or real)? How does the potential for increased data collection, storage, processing, and transmission reformat what is believed possible in the realm of determining what should and should not circulate and how? For decades, the automobility circuit has been augmented and managed through screening practices that insulate the driver from the road and from driving, replacing these mundane tasks with personal and profitable flows. Today these screening practices are increasingly attached to the mobile person as the noise of everyday life is reduced in order to maintain an ideally frictionless flow of personal (and generally profitable) "information."[9]

Flow: The Flux along the Circuit

Williams first described *flow* as the televisual techniques used to maintain audiences' attention to the television screen. Trying to understand how disparate television content was made to seamlessly flow together, he addressed a complex of societal, technological, economic, and cultural forces that had integrated broadcasting into a new way of living that was both more private and more mobile—what he called *mobile privatization.* Flow has been taken up in television studies for the past five decades[10] and has proven to be a productive way of thinking about the cultural impacts of technology in its focus on practices, techniques, and technologies of viewership. We are interested in the problematization of mobility and media at a specific juncture of change where users are newly orchestrating and being orchestrated by screens as they navigate through time, space, information, and entertainment. We are no longer fixed *on* transmission, but rather fixed *in* transmission. We are being tuned in, encircuited.

In his account of television, Williams focuses on how broadcasting—"a powerful new form of social integration and control"[11]—reorganized the movement of capital, subjects, and ideological content. Situating television in an expanding market of consumer durables such as large appliances and automobiles, Williams contrasts public technologies such as street lighting and railways with a group of emerging technologies and infrastructures that organized mobility, labor, privacy, consumption, technology, and the home in new ways as *mobile privatization.* In particular, televisions, automobiles, and suburbs were organized in conjunction with the needs of postwar industrial–consumer capital. Of importance to our analysis is Williams's explanation that these technologies helped to integrate the daily work commute with the suburban lifestyle, and how television's schedule worked to temporally and spatially organize idealized and highly normative ways of life.

Briefly, the broadcast schedules and spatial dispersion of television and radio networks and individual sets were an important part of the broader disciplinary structure of industrial capitalism during the mid-twentieth century. It is important to note that Williams's argument stood out from other British cultural studies perspectives at the time in that it was markedly (though not explicitly) archaeological. It focused on material as culture. Broadcasting media, then, weren't only ideological mechanisms; they were spatiotemporal machines for organizing circulations and flows of specific populations in space and time, through their daily routines. The two primary forms of broadcast were, of course, radio and television. By the postwar period, which saw automobile ownership and suburbanization pushed as nationalist ideals, broadcast radio had established a rhythm for keeping people moving between points of production, consumption, leisure, and rest. Media-consumption practices also became more private, even at the level of different media for different members of the family. For example, FM radio sets showed up in the bedrooms of teenagers beginning in the 1960s as youth became a problematic category uninterested in broadcast family TV fare.

The television schedule in North America from the 1950s to the 1980s (at which point cable begins its spatiotemporal disruption) provides a most exemplary means for thinking about how media work to organize temporality. The broadcast television schedule worked hand in hand with the highly gendered, raced, and classed organization of social expectations through a television regimen that coincided with specific periods for children's, women's, men's, and families' viewing. Television functioned to fill non-school time for kids, accompany women during their domestic routines,[12] address the masculine public with news after the nine-to-five of industrial capitalist production, and entertain the nuclear family prior to bedtime. On the weekend, children were entertained all morning long on Saturday, after which the masculine enterprises of sports took over. Sunday morning was often filled with religious programming, and the afternoon again with sports. Sunday evening, the family came back together for a final respite before the long work/school week began. The television season, with new shows appearing in early fall and finishing around May, followed that of the agriculture-based school year. The rerun season of summer was a time to be outdoors, to be away from home, and to be outside the seasonal flow of capital accumulation and knowledge accumulation (school). Later in the evening, television went off the air and each station on the dial was awash with staticky noise. In this way, the television regimen mirrored and enhanced the logistical regularity of industrial production and social reproduction. The televisual was a conduit for larger social and economic circuits, conducting the flow of bodies and capital through a home-centered means for making "idle" time productive[13] through "the work of watching."[14]

The rigidity of this schedule corresponded to the rigidity of a disciplinary society in which media systems, like most other social and economic institutions, worked at producing discrete circuits meant to regulate conduct according to a set of established norms. The suburban commute set the temporal organization of radio's drive time as it filled "dead time," provided valuable traffic and weather updates, and helped facilitate "water cooler conversation." By the 1960s, radio played a special role as a kind of "circuit breaker" that attempted to diminish the negative effects of the traffic jam by alleviating the individual effects of the increased number of traffic jams by sending drivers on less-congested routes or to collectively diminish their existence through extensive traffic updates. Radio stations began to use helicopters to aid in providing an aerial perspective. Radio provided specific kinds of ordering for highly feminized secretarial laborers listening to lite rock and pop interspersed with commercials for local shops; for masculine modes of production on remote construction sites where the rock stations on Friday afternoon would remind laborers that in the end they were "Working for the Weekend"; while the struggles of truckers against police surveillance were glorified by C. W. McCall's number one hit "Convoy" as they drove the endless highways;[15] and Muzak played across the mallscapes of North America, soothing the nerves of the perpetually indebted.[16] Radio as a mode of broadcast provided temporal and population specificity across the dial in ways that genre conventions reified "structures of feeling"[17] and programming conventions accommodated a "whole way of life."[18]

In the home, the role of broadcast media was to fit a different set of specific temporal orders in which the ideal television subject sat in a single place unmoving for hours on end as the flow of programming worked its magic. The couch potato is not a problem for broadcasters—in fact, quite the opposite. The couch potato is one element in broadcasting's most compelling circuit: a couch, a staring subject, and a television set—three objects fixed in space across an indefinite amount of time. Like a tractor beam, the audiovisual signals reach across the "family room" and draw the powerless human into larger economic, political, and social circuits. The threshold of the home ceases to serve its critical function when visitors come through the walls.

The fixity of broadcast's spatiotemporal ordering begins to ebb with the advent of cable, which rearranges television according to a narrowcast model and breaks up the rigidity of the schedule: cartoons can be watched at any time; sports are no longer isolated to weekend afternoons; domestic arts programming becomes a 24/7 option. The VCR allows for time and space shifting as programs can be watched at any time, over and over again, and the VHS tape can even be moved from one house to another. DVDs compress storage capacity and enlarge fidelity. Their lightness will usher in Netflix home delivery and replace the automobile-Blockbuster-home circuit with the internet-website-mailbox circuit. Soon after, broadcast and cable television recede as streaming establishes a ubiquitous televisual circuit. Audiovisual data become available across all of time and space. Everywhere at all times, everything can be watched. The spatiotemporal fixity of broadcast is permanently undone through the digital fluidity of 5G. Where a disciplinary order was supported through a discourse network guided by rigid divisions and schedules, the economic imperatives of innovation, destruction, speed, information, precarity, globalization, financialization, and privatization are leveraged and scaffolded upon digital circulation, storage, and processing.

Flow 1.0

Williams proposed the concepts of flow and "flow analysis" as a fluid replacement for static distribution analysis. Flow analysis worked to help a critic look beyond ratios of programming by type and commercials on a given channel for a given time, and instead focus on the ways in which sequenced programming worked to keep viewers glued to the screen and/or manage how they moved through their day in time and space. When conceived broadly, the concept of flow has the potential to shift our thinking about mobility and media as a loosely articulated movement through and connection to mobile, informational, economic, and political circuits. An expanded notion of flow draws attention to the ways that subjects are simultaneously active in a number of circuits that reorganize how life is conducted, managed, processed, stored, and digitized in real time across space.

Options for communicating and consuming have been vastly expanded and complicated by sensors, cloud processing, and big data[19] and must be taken into account when rethinking the more complex circuits that hold together mobility, consumption, work, and entertainment. As a method scalable to different levels of analysis, flow offers a flexibility that can be rethought in terms of sample size, duration, devices, and data type. This scalability can be leveraged in two important ways. The first is understanding the flow of circuits and persons at a broad level to get a sense of how populations are encouraged to activate various circuits throughout the day (the computer at work, smartphone in transit, and the television/smartphone/tablet and other media combinations for multitasking leisure time). The second is recognizing the potential value of close analyses that track the more nuanced connections to circuits: the news story on the TV that warrants a check-in for more information from Alexa, the work email that buzzes in your pocket as you sit down to dinner (it will only take a minute), or the ability to order a pizza without talking to anyone, from literally anywhere. Further, the seemingly mundane intersects with the global and political when activism (#MeToo, #BlackLivesMatter, #climatestrike) and nuclear diplomacy ("I too have a Nuclear Button, but it is a much bigger & more powerful one than his") play themselves out on the same model iPhone.[20] The activation of these circuits opens up various flows to and from nearly infinite points and are scalable from the single user to overall network traffic.

Consider the changes that have taken place in the past fifty years. In the North American broadcasting model, users bought a television to tune in to broadcasts where content was delivered for free and supported by advertising—a viewer bought a "set" to tune in with. This shift corresponds to a movement away from fixed "set" delivery eyeballs to advertisers during leisure. Today's user is required to keep current with an ever-shifting set of devices, maintain a combination of cellular and land-based connectivities, subscribe to an array of content streams, and manage personal libraries of digital products. These users are additionally swept up in generating lucrative streams of data that can then be variously packaged, aggregated, and resold on the broader data markets. Our advertisements have become targeted to the point that we know we are being watched by our screens and their attendant sensors, transmitters, data banks, and networks. We are also increasingly aware of the role our digital traces play in creating this encircuited experience.

The dominance of the broadcasting model has long since passed, but the functions of broadcasting continue to underlie the present form: we are informed, entertained, kept awake, distracted, and managed through and by screens. The "on-demand" model of today still casts a net over users, but their power to choose when, where, and to what they would be connected represents an important shift. Subjects are able to leverage screening technologies in order to navigate the moves from orbit to orbit. The standard home-to-work and work-to-home commute has been transformed for many into some combination of working from home and attending to home from work via screens, sensors, and circuits. Leisure time, travel time, work time, and domestic time can be at any time and in any space so long as those places and responsibilities are networked, or made accessible by the screen. Subjects are responsible for managing their flow, designing their own channels, stations, and content queues from the same devices they use to network with friends, family, and work while managing schedules that transcend traditional fixed nine-to-five work, rest, sleep, and repeat schedules.

But placing the subject at the center of this management doesn't rightly describe what is happening, and when taking recent developments into account it starts to become clear that the subject is just as much managed as it manages, and is accessed as much as it accesses. The user becomes a subject on demand, a database compiled into innumerable dividuals by advertisers, content companies, governments, and NGOs. As we engage with the circuit, it engages us in our continual engagement, a feedback loop of our various selves. While it is still in the service of the stability of traditional institutions (work, home, family, school, market, government, etc.), we see the activation of new forms supported and reinforced by the circuit in hyperpersonalized cross-pollinating arcs. Each arc (or segment) has the potential to arc (discharge energy to bridge the gap between two points) in order to generate something new. A dividual's activity is simultaneously made up of this constant suggestion of the same (Waze knowing that when you pick up your phone at a certain time of the day it's time for your daily commute to "Work") even as it provides you an up to the minute prediction of what will happen if you embark on a new path.

Screening Technologies

Williams's *mobile privatization* described a period in which space, time, and capital came to be organized by suburbanization, commuting, broadcast, and bureaucratic corporate industrialization. Our greatly altered mediascape demands a rethinking of what might help us understand a much different arrangement of living, moving, communicating, consuming, and being watched. We propose *screening technologies* as a specific set of technologies and practices that manage circuits of mobile conduct, consumption, and communication. Devices such as smartphones, wearables, and automobiles perform similar functions to practices for screening human and nonhuman mobile bodies to manage appropriate conduct. A circuit of user data for information that becomes the stuff of large-scale screening makes possible vast monitoring and subsequent redirection of conduct.

Screens are everywhere. More to the point, the kind of screens most associate with the term is far too narrow to encompass our analysis. "Screen time" has become a household phrase, with adults averaging around eight hours of such time in 2018 (10.5 hours when radio and internet-connected devices are included).[21] We are not only aware that we are connected, but willingly submit vast amounts of data about our location, preferences, likes and dislikes, routes traveled, heart rate, steps taken, and more to unknown third parties through these and other connected devices. We manage personal and professional voice calls, email and text, social media, health, finances, purchases, games, navigation, and life more broadly through screens and screening. Sometimes we know those ads are targeted *to us specifically,* reminding us that someone or something is watching, processing, and remembering us.

Mechanical screens, best exemplified by the colander or sieve, physically allow some substances to pass while denying access to others through channels (holes) of fixed size and shape. They act as rigid systems of normalization and separation (as in disciplinary power).[22] Mechanical sieve analysis separates and distributes objects according to a fixed set of characteristics (size, in the most obvious cases, where civil engineers assess the makeup of a composite such as concrete). Population dispersion that acts according to *fixed* markers[23] of difference, such as race, gender, or age, can be thought of in analogous terms as kinds of screens. Digital and biometric screens work in more flexible ways: their sieves can be altered one individual, place, or metric on demand. Which information, through what channel, at what speed, and for how long can easily be regulated by simple and prevalent software. The analog screening technologies of broadcast that Williams described acted to normalize through few channels according to fixed schedules in limited spaces. By contrast, televisual, automobile, and communicative circuits today are interactive, flexible, and ubiquitous. Their flows change shape, direction, and speed to suit a multitude of needs, though one seems to loom above the others: capture. In the sensor society, "devices and applications developed for one purpose generate information that can be repurposed indefinitely,"[24] and the possibility of collection is reason enough to siphon as much data as possible. Everything that can become a data point has become a data point, even if no one is sure what to do with it, yet.

In their 2010 article "From Windscreen to Widescreen: Screening Technologies and Mobile Communication," Jeremy Packer and Kathleen Oswald identified six necessary characteristics of screening technologies: storage and access, interactivity, mobility, control, informationalization, and convergence/translation.[25] *Storage* is best understood as a general movement from local to networked storage, with physical media giving way to the always accessible cloud. *Interactivity* may be better understood as *connectivity,* necessitated due to a move from possession to access. Rounding out the first half of our list, *mobility* has been vastly altered in what we can access and where, and in turn who can access us: we not only do everything everywhere, but our always-on connection makes possible a virtual reconstruction of what, when, where, and how we live our lives. These characteristics constitute the foundation of the mobile communicative circuit.

The next three characteristics describe the operating logic of screening technologies: control, informationalization, and convergence. *Control* leverages storage, connectivity, and mobility to facilitate constant activation of the circuit, connecting users to data and user data to databases: as users activate networks they open themselves to mechanisms through which they are filtered, sorted, and screened. These data are then compiled and acted upon algorithmically through *informationalization,* whereby past activity and patterns of behavior are brought to bear on the present (and predict the future). *Convergence* is realized in the present mediascape in what Vincent Mosco describes as the "Next Internet" composed of ubiquitous cloud computing, big data analytics, and the Internet of Things (IoT).[26] As more and more of the world is known to sensors, networked, and made meaningful in real-time interactions, screening technologies digitize, measure, and aggregate us as individuals and collectivities for biopolitical action. The desire for this level of population control is most certainly not new, though the realization of this networked mechanism of real-time control from a distance is made possible through a reduction of everything to a digital code that can be stored, accessed, and acted upon by machine networks.[27]

Starting with the production and expansion of automobile screening, we focus on methods of screening out unwanted elements and the eventual tuning in of the driving experience and emerging efforts to tune driving back out in favor of more profitable circuits. The role of digital screens is fundamental in enabling increased spatial and temporal fluidity of population, information, capital, and labor to ultimately enable greater social, political, and economic control through a seemingly contradictory extension of self-empowerment. We next review screening technologies by asking how screens function in the automobile, suggesting that they are first used in the act of driving, then for entertainment, and finally to mixed-use screens that blend a number of driver information functions, communication services, and features that operate both inside and outside the car.

From Windscreen to Widescreen: From Sensation to Information

The first cars were not much more than an engine, four wheels, a steering wheel, and a chair for a driver. This left drivers fully exposed, but also highly tuned in to the sensorial effects of high-speed motion. Eyes, ears, noses, skin, and even mouths were buttressed against a heightened level of sensorial input that increased along with speed. Bodies collided with signals at a pace that was alarming at the time, producing full-spectrum noise. Initial screening technologies were developed to diminish and focus stimuli to create a comfortable space in which the driver is able to process "relevant" information in order to safely operate the vehicle. Road noise (doors, windows, a roof, and eventually insulation), cold (heater), and ruts and cobblestones (shocks, inflated tires) all had to be "tuned out" in order to increase the fidelity of useful data (oncoming traffic, curve in the road, darting deer, oblivious pedestrian). Additional developments continued to enhance the function of the primary screening technology. Wipers kept the windshield clear of raindrops and clean of bug splatter, road grime, and salt residue. Safety glass reduced the chances of shrapnel in the case of impact-related windshield failure. While the sun provides light to navigate, too much creates visual noise and was attenuated with visors, sunglasses, and, many years later, tinting.

While the windshield remains the primary screen through which visual information is gathered in an automobile, instrumentation keeps the driver in touch with the vehicle and other elements of the driving environment. Though the dashboard originally served the basic purpose of reducing driver contact with the road (in a very material sense), today's dashboards house instrument panels where what happens outside of the driving cabin comes to be understood as information rather than sensation. With increased insulation from the outside environment, drivers can no longer sense outside air temperature, let alone engine temperature. It is difficult to hear the engine and know when to shift, so a tachometer becomes necessary. Speed itself is no longer something that can be sensed, and the speedometer becomes increasingly important as automobiles develop from a novel technology to a widespread mobility system that requires the control and governance of mobile conduct.

With the emergence of electronic instrument clusters in the 1980s, new calculations could be performed, enabling new analytics for the driver. Significantly, something as simple as a miles-to-refueling calculation is not possible until microelectronics enable the storage and access requirements for real-time calculations that take into account fuel efficiency, fuel on hand, and speed to produce information. The car, as with other media, eventually moves from analog measurement and display to becoming a digital information producer, distributor, and processor. This shift corresponds with new technological capabilities and a further removal of the driver and mechanic from the phenomenological material into the seemingly immaterial realm of the digital.[28]

Highway Hi-Fi, or Making Music Auto-mobile

Increasingly sheltered from the elements, drivers and passengers opted for mobile entertainment. From the first portable radio in 1929—the "Motorola"[29]—it was clear that mobility would have a soundtrack. The automobile provided not only the power and space to move the radio but the electricity needed to power it. As tubes gave way to transistors and analog tuning became digital, greater capacity for flexibility and control arose. First drivers selected broadcasts to tune in to, next drivers were able to make personal libraries mobile, and today the car has become a site of tuning in to our libraries via data streams. While mobile record playback was not a success, the 8-track was specifically designed for mobile listening and enjoyed a good decade of mobile music supremacy before being replaced by the cassette. A standard feature by the 1980s, the cassette enabled data transfer/translation from any previous format (radio, LP, 8-track, CD) to a mobile-friendly platform. Cassettes could move from place to place, person to person, mix media from various sources, and be copied and shared without damage to the original. The cassette initiated a large-scale reorientation for mobile media: recording, mixing, producing, reproducing, format shifting, time shifting, distributing, and listening become a more mobile process, and one that is as accessible to the masses as the copier and the VHS tape—a new standard for mobile privatization. Like their counterparts in publishing and film, music industry executives worried that commercially produced music would no longer be profitable in light of these new abilities, though commercially produced cassettes soon became the medium of choice.

This dominance did not last long. The shift to digital was introduced via the compact disc, not an addendum but a vinyl replacement that promised "perfect sound forever." The format brought about greater content control, storage capacity, and portability, and, taken together with file sharing, paved the way for the now ubiquitous MP3 format of compression. A mere decade ago the need to store your MP3 files on something portable like an iPod Shuffle, which made a stack of audiophile LPs seem obscenely excessive to the highly mobile or space-conscious,[30] was still necessary to move audio data. Today, storage capacity is obsolete if some device near you is within reach of a digital network; holler to a digital assistant, and Spotify's library awaits. It doesn't get any lighter than that.[31] Access, connectivity, and mobility have been fully actualized in the cloud-based services of today, enabling new forms of control, informationalization, and convergence across devices and spaces. Mobile communication may have been nurtured in the automobile environment, but it has since become detached from the automobile and attached to the person. The automobile is no longer a necessary power source and storage device; it is a giant portable speaker and a conduit to libraries stashed elsewhere. The car is a screening device among other screening devices, of a kind with the desktop computer, the tablet, and the smartphone—a communication device that happens to carry your body across space and time, as well.

Moving without Traveling

As electronics and eventually computers come to be integrated by manufacturers, different information is collected (and stored) within the car, and features such as anti-lock braking systems (ABS) and traction control systems (TCS) begin to make the car less of a machine and more of a device. Its maintenance, then, is likewise increasingly handled by technicians rather than the mechanics of yesteryear. The car becomes an information space, where data such as engine temperature and speed are subsumed by fuel efficiency and EcoDrive scores. Milestones, road signage, and maps give way to GPS software that simultaneously guides you, the driver, and shows the world where you are, everything in your reach, locations rushing around you. As the driver is increasingly abstracted from driving and wayfinding, automobility becomes more about passing time in relative comfort than operating a complex machine.

While music has long had a place in the car, the road always served as the only real-time visual. Signaling and signage aside, advertising[32] and architectural sites such as Sunset Boulevard in Los Angeles[33] or the Las Vegas Strip[34] were new media phenomena of a motor age. Stopping the vehicle, however, afforded new visual sensations such as the drive-in theater, where the car became a living room and the windshield acted as a screen for entertainment rather than navigation. At the same time as the car entered the cinema assemblage of the drive-in, Lynn Spigel identifies a shift from the era of the "home theater" in the 1950s to the era of the "mobile home" in the 1960s, explaining that the portable television "promised to take the interior world outdoors" in what she calls "privatized mobility."[35] Spigel argues that "privatized mobility" demonstrates a shift toward seeing the home as "a mode of transport in and of itself that allowed people to take private life outdoors" rather than bring the outside in.[36] The increased density of circuits has led to new possibilities in storage, connectivity, and mobility that make the car a primary site for bringing private life everywhere through screening technologies.

From the handheld televisions of the 1960s to the VCR as a high-end mobile entertainment option of the 1980s and 1990s, the 2000s saw a shift to portable DVD players and finally a more or less complete transition to screens with no external media. Portable gaming has moved from the battery-thirsty Game Boy and Game Gear of the 1990s to the ubiquitous screen, with much of gaming moving to smartphones and tablets. Much like the portable radio in the bedroom of the teenager in the 1960s, drivers and passengers have their own private media circuit and are insulated and isolated from both the world rushing by and the family members sitting next to them. Journeys can be augmented, accompanied by, or in fact replaced with televisual journeys. Pocket technologies are transforming passengers into packets: the family unit dematerialized in transport, only to be rematerialized as family at the destination. Transportation becomes teleportation as the experience of travel is displaced by a multitude of individually mediated televisual journeys.

The Automobile Circuit

When Williams first conceptualized mobile privatization, the automobile and broadcasting were oriented to the spaces of work, home, and nation. This communicative network was well circumscribed, and the automobile played a straightforward role. Contemporary drivers, however, are at the nexus of a more complex and fluid set of relations. At least a triple displacement is at work in the automobile network: first, the driver is distanced from the road environment; second, the driver replaces the vacuum of sensation with entertainment and advanced information systems; and third, these networks displace the driver from driving itself. We next look at communication from and within the vehicle as steps toward this third remove: autonomous driving technology, the telos of the circuit.

As drivers began to fill roads in the early 1900s, a new crisis of mobility was at hand: How should movement be managed as more vehicles are entering already busy thoroughfares?[37] Numerous media were used to assist inter-driver communication: signs, turn indicators, traffic lights, horns, mechanisms for overcoming the noise introduced by engines and roads, and lamps to light the way in the dark. Inter-vehicular voice communication was initially introduced via CB, and began to reorient and expand the automotive communications network. With the introduction of the mobile telephone, the calculus for interlinking home, production, consumption, and nation radically changed. Early car phones consisted of a radio transmitter and a large logic unit inside the trunk that enabled the car to connect to the telephone network, technology that remained car-bound until the microchip reduced them to a fraction of their former size, leading into the development of the cell phone. While the car phone was named for the place it was installed and used, the cell phone is named for the new configurations of space that infrastructure makes possible.[38]

Infrastructures of automobility have often been discussed in terms of interstates, freeways, turnpikes, collector and local roads, as well as fuel racks, rest stops, and gas stations. The fuel that drives most of the automobility system is still refined from oil that comes at a great environmental, economic, and political cost.[39] While these elements deserve much continued discussion and reflection, they are not the automobility infrastructure discussed here. Instead we focus on the communication networks within the vehicle: the miles of wiring and dozens of electronic control units that constitute the electronic infrastructure of an average automobile and extend out to the broader communication infrastructures that encircuit the automobile and driver.

The Copper Development Association explains that "today's luxury cars, on average, contain some 1,500 copper wires totaling about one mile in length, thanks to continuing improvements in electronics and the addition of power accessories. In 1948, the average family car contained only about 55 wires amounting to an average total length of 150 feet."[40] The association notes that advances in chip manufacturing means that chips may increasingly be made with copper rather than aluminum, enabling higher transistor density. Combined with electronic components and more than one hundred electronic control units (ECUs) in an average car,[41] a wiring harness and attached components looks like (and arguably is) the nervous system for a giant robot.

Whereas electronic components were responsible for about 1 percent of the total cost of a vehicle in 1950, that number jumped to 10 percent in 1980 and to 30 percent in 2010.[42] *EDN* journalist Carolyn Mathas explains the inclusion of ECUs was initially driven by government mandates for fuel economy, stringent emissions standards, and safety, but that increased passenger comfort and entertainment systems drive up cost to the extent that by 2030, half of the cost of manufacturing a vehicle will be the electronic components.[43] Mechanical systems are increasingly being replaced with electronic ones, such as drive-by-wire, electronic brakes, and fobs and apps that grant access over keys that unlock the cabin and turn over the engine.

Figure 4.2. Screenshot from "Wiring Harness Layout—How to?" by Allison Customs'—PROJECT CAR TV. Jeffrey Allison, Allison Customs, LLC, Jeffery Allison, Bloomfield, New Mexico.

A 2010 article in *Assembly Magazine* explained that the role of wire in drive by wire systems (x-by-wire) does not necessarily mean more wire.[44] At that time, Todd Hubing of Clemson University's International Center for Automotive Research explained that the use of digital data meant more information could be sent over less wire, which made sense in an environment with rising copper prices and falling electronics prices.[45] A 2019 report by Deloitte projected that while electronics and semiconductors made up approximately 18 percent of the cost of a car manufactured in 2000, they would constitute 40 percent of the cost by 2020 and 45 percent by 2030.[46] The vehicle has become more device than machine with its fine mesh of wires, sensors, and control units supporting a range of communication, entertainment, and driver safety and assist features. Some are outfitted with features sufficient for autonomous operation, or close enough "autopilot" capabilities.

While Windows Embedded Automotive (formerly Microsoft Auto) was decommissioned in 2021 (this was the foundation for Ford Sync and Blue&Me), a newer Azure cloud-based service called Microsoft Connected Vehicle has replaced it. Microsoft has also recently partnered with BMW, Mercedes, Volkswagen, and Ford for cloud-based services.[47] While the company has pulled the software from the vehicle itself with the demise of Microsoft Auto, the new economic model of collecting data to be stored and processed remotely before instructions are sent back out to the vehicle-as-terminal is part of the larger IoT/cloud/big data economic model.

New mobility concepts such as KMPG's Mobility as a Service (MaaS) index, which attempts to envision "policy and customer-optimised mobility services"[48] in light of increased congestion and technological solutions, shows the potential for widespread redefinition and regulation of public/private mobility systems managed by cloud-connected automobile circuits. A March 2018 Frost and Sullivan report suggests that a transition from driver to passenger could happen over the next ten years, as we have already reasonably moved halfway through the following progression to full automation: feet off, hands off, eyes off, attention off, passenger.[49] The report explains autonomous technology is facilitated by "sensor fusion" where data are gathered by sensors and understood in the context of other data in the "ecosystem." The report indicates that electric vehicles are a preferred base for autonomous driving and that these vehicles would ultimately be leveraged as rideshare platforms or through shared or exclusive licensing, rather than pure ownership models.

Importantly, the broader subject-based acceptance of contracting and leasing models is largely introduced by screening technologies. We no longer tune in to broadcasts; we subscribe to libraries. The smartphone has replaced the car as the first major expensive item a teen becomes responsible for, and as costs skyrocket there has been a general shift from device ownership to device leasing. Leasing, subscribing, and contracting are all circuit-based activities in that they require a constant verification (usually payment) in order to operate. Media and mobility are two strands in a copper wire bundle, intermittently connected by broader cultural shifts. In the present, they are coming back together in autonomous automobility.

Consider automobile leasing, where a driver puts down money "due at signing" and takes on a monthly rent for a vehicle (opposed to paying cash or making a down payment followed by a financed sum due over time). Leasing vehicles is on the rise: in 2008, 20 percent of new car sales were leases,[50] approximately 30 percent of all new vehicles bought from 2017 to the first quarter of 2020 were leases, falling to about 25 percent of new vehicle purchases through 2020 and the first two quarters of 2021.[51] One of the many described advantages of leasing is being able to "upgrade" to a new vehicle at the end of a term while avoiding the looming threat of negative equity (or "being upside down") in a financed vehicle at trade-in. This loss in value turns out to be an upside for the lessor, as the vehicle can be depreciated on their end for tax purposes. The global vehicle leasing market was valued at $69.27 billion in 2020 and is projected to grow to $123.87 billion by 2028.[52]

Ubiquitous computing, the IoT, and constant mobile connectivity have escalated the logic of the circuit, conditioning an increasing number of users to be comfortable with lease and subscription models. Today's average U.S. household subscribes to more than three video streaming services,[53] and well over half of them subscribe to Amazon Prime.[54] Add to these music subscriptions (Spotify, Apple Music, and Amazon Music Unlimited), software subscriptions (Office 365, Adobe Acrobat Pro), and vehicle software subscription services such as OnStar and similar manufacturer-specific vehicle apps with monthly fees. Even vehicles are increasingly available in a monthly subscription model, with a lower cost due at signing, a higher monthly fee, and the ability to cancel after a few months (rather than a couple years).

Many drivers want to be in newer models, as safety, fuel economy, and infotainment features are upgraded rapidly. Adaptive safety features in the vehicle are on the rise: precollision detection was available on 4 percent of vehicles in 2007, rising to 57 percent in 2017.[55] The National Highway Transportation Safety Agency (NHTSA) mandated in 2014 that by May 2018 all new vehicles in the United States had to come equipped with a backup camera and screen to reduce backup fatalities. Increasingly stringent emissions and fuel economy regulations (which have resulted in environmental improvements and increased circuitry in the car since the 1970s) are moving car makers to invest in electric and hybrid technology, though the recent downshift in fuel economy targets from 54 to 40 miles per gallon by 2026 has caused some innovation to stall.[56] Electric vehicles accounted for only 2 percent of all new cars sold in the United States in 2020, but in Europe that number was 10 percent.[57]

As Tesla's marketing makes abundantly clear, electric vehicles and autonomous vehicles go hand in hand. Edmunds reports: "With the future looking increasingly autonomous, electric powertrains are the natural accompaniment; they have fewer moving parts when compared to internal combustion engines and are simpler for computers to drive."[58] The NHTSA has several initiatives to push more automated safety features into vehicles as optional (and in some cases standard), and the U.S. Energy

Information Administration also advocates for autonomous mobility in a May 2018 report, including the image in Figure 4.3 illustrating slow erasure of the driver. The report explains some of the potential benefits of autonomous automobility: road safety and the reduction of fatal crashes, increased mobility for those who are disabled or too young or old to drive, and decreased congestion on roads. There is one more benefit, however:

> The average American driver spends 294 hours behind the wheel per year. In aggregate, the nation's 222 million drivers spend about 65 billion hours driving their vehicles. Autonomous vehicles could not only reduce traffic congestion and time spent on the road, but also provide time for drivers to engage in other activities.[59]

As is the case with most future hype, autonomous automobility promises that more and better technology and connectivity will solve our mobility problems from the individual to the global scale. The circuit model is motivated by the market, supported by those who would have access to and control over flows, and sustained by our collective participation and enthusiasm. Mobile music and communication-enabled spaces of production and consumption enter the car through our voices and ears. Autonomous cars will free our eyeballs from the road and place them back on screens where we can be producers, consumers, and managers of a profitable existence. We will flow from place to place in a seemingly frictionless fashion that is underpinned and made possible through vast material and digital infrastructures, service contracts, cloud computing, invisible labor, and remote energy generation. We will be productive; we will be comfortable; we will be entertained.

The exponential progression of encircuiting things, people, places, and transactions seemed to be inevitable and invisible until Covid-19. The pandemic has been a revelation of the circuit in more ways than one. We learned that much of life could continue while we stayed remote. Disciplinary logics of confinement operated to quarantine the sick, and governmental logics urged us to slow the spread, wear a mask, take a vaccine, to manage ourselves. The logic of the circuit kept us connected in our productive and consumptive orbits, and the vast infrastructure making that possible carried that weight through most of the first year of our new normal. Part of that new normal is the microchip shortage precipitated by the pandemic, a material vulnerability that betrays the apparent seamlessness of circuit

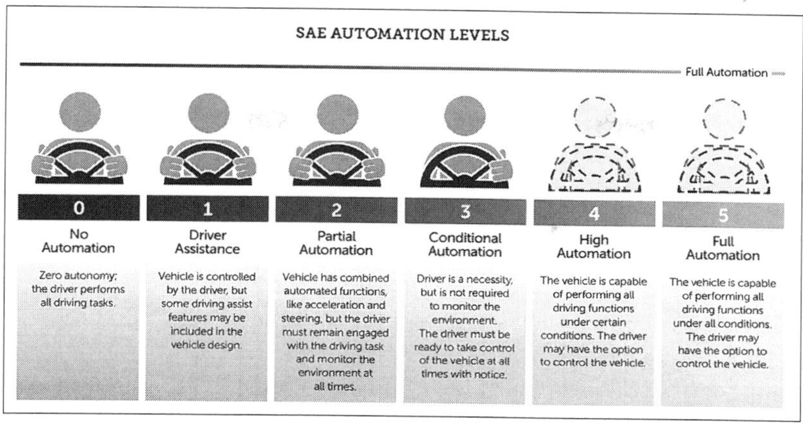

Figure 4.3. Automated Driving Systems 2.0, A Vision for Safety. Image from the U.S. Department of Transportation, NHTSA.

society. Critically, the shortage hinges on a need for semiconductors, the building blocks of encircuited life. Their scarcity not only highlights their importance in electronics and traditional consumer durables such as home appliances and automobiles; more broadly, it highlights the scales at which circuits and circulations are fundamental. The failure to manage viral circulation disrupted capital, cultural, social, economic, and human circulations, which in turn weakened the capacity to create the fundamental building blocks of our prison house. The degree to which these failures and new governmental problematizations generate new forms of encircuiting is an ongoing development.

Conclusion: Circuit Breakers, Braking Circuits

Importantly, our explanation of screening technologies is not only a matter of control, but of encircuiting: a form of guidance that is empowered, exerted, and expanded by, on, and through each node; each privately or publicly mobile user engorges the circuit's capacity. Perhaps what is most pressing is to create new ways to screen what or who is screening us, to understand the nature of the relations that the newly encircuited spaces of the automobile enable and attract. This is of critical import: if the automobile becomes an open site, then it can be taken over externally. Imagine bumper cars when the juice stops flowing and the ride is over. Or Khan's wrath when, after successfully commandeering a Starfleet science vessel, Kirk and Spock seize control of the bridge remotely, lowering his shields and blowing him to bits. Imagine, for a moment, your car going Christine, possessed by an unknown and evil force from beyond.

Our history is one in which automobiles have become increasingly and paradoxically individualized yet externally dependent. They require bandwidth. They simultaneously travel the highway circuit and the "information superhighway," as it was once called. The driver was historically faced with an increasing number of screens that dissolved the immediate collision of mechanical forces and instead plugged humans into electrical circuits. At first, mechanical devices like windshields, shock absorbers, inflatable tires, heaters, fans, safety belts, and sun visors encircled drivers in a space shielded from wind, rain, cold, heat, bumps, UV rays, rocks, and driver expulsion and were seen as solutions to manage the interface of human, machine, and rapid movement. Dials and knobs were mechanically attached to armatures, cranks, levers, rotors, and pulleys. Analog media systems modulated radio signals or read magnetized tape. Currently, mechanical buttons and dials, as well as screen-based "buttons," don't actuate a lever, but produce a signal for the circuit. The mechanical remnants of steering wheels, brake pedals, and gear shifters are no longer needed per se. Rather, signals simply need to be humanly or algorithmically inputted into the circuitry. Instrument clusters are replaced by voice-operated menus; a box of cassettes is replaced by a playlist; a book on tape becomes a podcast; GPS begets Internet access, and the automobile itself is fully encircuited. Just another screening technology; an informationalized, interactive, and converged digital mobile habitat whose use is monitored and governed through screens built into expansive social and economic circuits. Cars have added storage capacity not only due to their increased size, but due to digitization and being networked. Drivers, passengers, luggage, and data are all stored and transported. Yet, screening technologies are not only present in the automobile. Screening technologies are the portals by which mobility and immobility increasingly come into being. Money, bodies, goods, noise, culture, and cars are but a few of the forms organized and given direction by screening technologies. Mobile communication is in a way a redundancy in terminology. Couldn't either term suffice? But as long as scholars work to make sense of "it" we suggest orienting such scholarship in terms of the circuits that organize mobile thought and conduct.

While the logic of the circuit in automobility has been largely about mating material circulation with information management, arguably the perfect mobile circuit would involve not only no driving but no physical transportation at all. The heft of material existence carries with it too high a cost in fuel. Rather, as with everything else, turn bodies into information, beam it at light speed, reassemble elsewhere. The car, like the chariot, takes up space at a museum, but nowhere else. The unique moment of material potential and social demands pass. The automobile is obsolesced like the Ginsu knife and the 8-track. Its remainders appear on eBay and in the back of junk stores devoted to a "vintage" that is wholly undesirable. But if this advancement proves to be beyond human capacity, maybe something else will break the current circuit.

Ubiquitous computing and constant connectivity are at the core of our media and mobility landscapes. Wherever you go, there you are—on your screen, a data point moving through time and space in physical and virtual places. The only sure way to elude surveillance is to avoid the very circuits that make contemporary connected life possible: give up your cell phone, steer clear of license plate capture on toll roads and public streets, avoid facial-recognition surveillance, log in nowhere, eschew credit cards and checks, don't carry identification. Leave no trace.

The logic of the circuit and practices of encircuiting are the very stuff of society. Our government, military, economy, health-care systems, communication networks, and education infrastructure, to name a few examples, rely intensively on both circuit logic and circuits themselves to function even at the most basic levels. Data must flow. The marriage of the material to the informational will never be undone, despite the risks such connectivity poses in opening infrastructures as critical as energy, finance, transportation, government, and health care to espionage, manipulation, and attack. The upside is simply too great to risk missing out on the gains afforded by the synergistic effects on encircuiting.

Natural disasters such as earthquakes and hurricanes bring the centrality of circuits into sharp relief: where is everyone, what is their status, what is and is not working, how do we bring things "back online" and make them knowable so that they can be fixable? War games that model cyberattacks serve as crisis preparation for worst-case scenarios such as takeover and sabotage of critical systems. Our encircuited world brings with it the inevitability of its accident, cascading systems failure. Just as the weakness in one organ system can lead to the decline of others until the body as a whole fails.

This fragility is clearest in a *sudden* crisis, but we can also see it in a shortage of microchips, toilet paper, and used cars; in supply-chain disruptions, rolling brownouts, or an exhausted and reluctant workforce. It is hard to know for sure if the system is failing without the blockbuster imagery of cars wrecking on the highway, reactors going critical, or an EMP that renders everyone's electronics useless. The cascading failure of the circuit could very well be playing out now in *slow motion.* You may find it in sparse car lots, in irreparable devices, and empty spaces on grocery shelves. The circuit may fail because our human systems, including the bodies of workers across the globe, simply can't keep up with the demands of encircuiting for encircuiting's sake.

5

How We All Were Committed

Automating Medial Madness from Eyeglasses to Google Glass

> *Every human eye has a blind spot near the center of the visual field. This is not about peripheral vision or a view from the margins. It is right at the center of experience. The eye does not know its own blind spot, mistakes it for vision. Nor does the mind's eye and so gives meaning to what is not really there.*
>
> *Every human brain fills in what is missing, blinding each of us to our own blind spot. A human paradox, each of us sees where we cannot and do not. The mind is positive that it sees what is really there. The arrogance of the human mind, of the human eye is rooted firmly in physiology. Of this personal vision, a worldview is born and an entire life lived. Yet the center does not hold because it does not even exist. Right in the middle of each person's universe is a big dark chasm into which each of us must inevitably tumble again and again unnoticed and unnoticing.*
>
> Ellyn Kaschak, "Believing Is Seeing: The Blindness of the Sighted"

Self-enucleation, defined as any self-inflicted wound that permanently severs the eyeball from its socket, is a rather grim solution to preventing unwanted visual stimuli. It is also referred to as oedipism, in reference to King Oedipus's inability to *properly see* his father and mother, which ultimately led him to extract his own eyeballs. Self-enucleation is rare, but the most common causes are schizophrenia and drug use,[1] both of which can be accompanied by auditory and visual hallucinations. One case involved a forty-year-old man admitted to a psychiatric ward.[2] He referred to the eye of Horus tattooed on his left palm as his "evil eye."[3] He had violently removed his actual right eye, and swallowed it because "he didn't want to lose it."[4] He later attempted to complete a bilateral self-enucleation by tearing his left eyeball out with his fingers. "He sublexed the globe but did not complete the removal."[5] He remained under heavy sedation and careful watch.

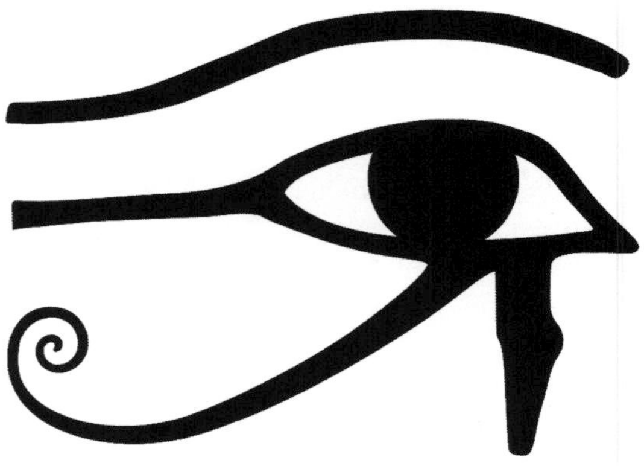

Figure 5.1. The eye of Horus. Created by User eff Dahl / CC BY-SA 4.0 (via Wikimedia Commons).

While this case may seem particularly grisly, scholarly articles in ophthalmology and psychiatry journals are filled with brutal accounts and accompanying illustrations of the tools and effects of such self-mutilation. And while it is in no way the case that the subjects of these horrific acts are all driven by a desire to stop visual hallucinations, it is more clear that there is a relationship between visual perception and schizophrenia.[6] Seeing something that is not present is considered one of the primary indicators of schizophrenia,[7] and one sufferer explained that "hallucinations really effectively kidnap the senses."[8] Further, the neurophysical interrelationship between one's eyes, one's brain, and hallucinations is complex; hallucinations result from numerous physical, psychological, neurological, and pharmaceutical stimuli.[9]

Human vision is likewise complex and highlights some of the ontological and epistemological contradictions bridging reality and perception. Vision emerges through external and internal processes. The manipulation of human vision has historically been a concern of political philosophers,[10] media industries, and ophthalmological technicians. Attempts to control what can be seen have ancient roots that are grounded in institutional governance and ethical regimes of self-control. The question we address here is: What happens when technological enhancements to vision, from eyeglasses several centuries ago to developments in digitally enhanced forms of visioning glasses of the past decade, are used to normalize social expectations in the areas of labor and sociality by altering vision to such a degree that it mimics a hallucination? As sight has been preprocessed through eyeglasses for several centuries, cultural sentiment tends *not* to think of the resultant vision as a hallucination. The hallucinatory results of Google Glass or Microsoft HoloLens may instead bring the technologically processed dimension of psychotic vision into clearer focus.

Recent research addressing commonplace visual hallucinations suggests that when what a person actually sees is contradicted by what a person expects to see, a hallucination can occur. More technically, "Psychotic experiences arise from an increased use of prior knowledge in constructing meaningful percepts from ambiguous sensory inputs."[11] Human vision arises from the intersection of two circuits. When the "bottom-up"[12] circuitry, cortical memory, overpowers and short-circuits "top-down signals,"[13] the light-generated impulses coursing through the optic nerve, a hallucination occurs. From a media-theoretical perspective we could approach this as a problem arising out of the inability to properly process newly inputted data due to an overemphasis upon previously stored data. New data aren't allowed to speak their truth, because previous knowledge overdetermines the capacity to make sense of the new. Signals are sent from two directions; two circuits carry electrical impulses. One is triggered by light waves penetrating the eye; the second is triggered by visual memories stored in a brain that was wired to process visual data millennia before it was wired to think critically about what it was seeing. What happens when a third circuit bisects the other two and introduces purposeful noise into the system? What happens when individuated vision is preprocessed in real time? What kind of psychosis might result from machinic hallucinations? Google's neural network for image recognition gives us some indication.

The alteration of human vision by technical means has a long history, which we will partially impart over the following pages. More recently, though, the capacity to process and manipulate visual data in real time from the same data perceived by human vision is being used to create another kind of hallucination driven by a wide-ranging set of lived contradictions, institutionally organized agendas, and psycho-scientific epistemologies of subject formation and normative sociality. This form of hallucination is only one of the two interrelated ways we want to investigate how we might imagine alterations in visual perception as driven by medial madness. At the heart of attempts to use Glass for rehabilitation lies a series of psychological models and theories for how to integrate Glass subjects into the social order. So we ask a simple set of questions: How do modes of visual problematization assume a scientifically determined

Figure 5.2. Psychedelic image created with Google AI. Licensed by Google Inc. under a CC BY 4.0 license.

normative form of vision? What is the history of this normativity? And what are some of the disruptive moments that highlight the expectation of "corrective" vision that are magnified by something akin to Google Glass?

We rethink visuality not in terms of how cultural expectations alter or create the meaningful modes of signification associated with what humans biologically collect as visual data, but rather as always already a series of instances of processing light via mechanical/technical means.[14] We are first and foremost interested in how light is processed to make the world navigable and relatable to human needs. If the need is a solution to a problem, then sight becomes a mechanism of problem solving, a way of collecting relevant data necessary to enable human collision with space. Any light processing that benefits humans in such collisions might be considered a success. Such light processing has primarily been a biological process to this point in human history. Increasingly, it will become a technical process: a mechanical process of bending, sorting, storing, and processing light rays into systems of reproducible and dependable information that is stored in a remotely accessible data bank.

This chapter will address these issues via a genealogical investigation into the role of eyeglasses, both the analog and digital varieties, in five distinct circuits to determine the effect of eyeglass media in their respective systems as a tool for managing the self in accordance with discourse networks. The first half of the chapter examines the origins and development of reading glasses in the late 1200s. These glasses emerged for those with presbyopia (farsightedness) in response to increasing concerns about managing and prolonging vision in order to maintain the necessary labor associated with reproducing religious documents by hand in order to extend the information machinery of the Catholic Church. Analog eyeglasses, or spectacles,[15] demonstrate the working relationship of human and media in emergent governance, whereby the human is shaped by the medium at the behest of a governing body. The second half of the chapter considers the use of Google Glass as a mode of realigning the visual field to account for other kinds of medical and psychological problematizations, such as autism and Alzheimer's. Here, the visual becomes the means by which "faulty" internal processing can be brought into alignment with social norms by overprocessing the visual data before it reaches the wearer's eyes. This sort of preprocessing is very similar to the normative alignment of spectacles, but of course it is also much more powerful. It does not only bend light; it uses light to access a vast database which can then repurpose and represent that data in novel and highly specific ways. Our various forms of problematization (cognitive behavioral therapy, memory stimulus, memory augmentation) operate as meta-problematics that run through a more technical set of problematizations in which the visual can provide valuable data for the interconnection between human perceptual capacity and social demands. Re-seeing or pre-seeing the world, Glass provides visual data necessary for allowing the wearer to recognize the patterns of emotional call and response deemed socially acceptable. Light control becomes a highly disciplinary mechanism under the aegis of a therapeutic regime.

The development of spectacles cannot be isolated to an Archimedes or Daedalus figure alone in a workshop. Instead, a network of practices, media, institutions, and technologies were responsible for the development of the information-processing technology. This chapter considers the ways in which spectacles were developed due to the antagonistic relationships between technologies and the senses and between technologies and other technologies—an escalating competition urged on by the will to know and the structural organization of knowledge through media—and it further investigates the power relations embedded in the technology that serve to manage the self. Glasses function "as ontological instruments intended for correcting the dysfunction of the experience of vision."[16] This discussion invokes a Kittlerian frame of media escalation, where media are always working within the limits of the sensorial and corporeal bandwidth of human sensation. It also clarifies McLuhan's adage that "It is only too typical that the 'content' of any medium blinds us to the character of the medium."[17] Because our concerns are often focused on print or digital media, we fail to recognize the medium of glasses in shaping our experience with content. To clearly understand and discuss the role of the spectacles in the optical mechanism, it is important to briefly examine the unaided human optical mechanism.

Optical Mechanism

The visual process has been considered in diverse ways, as illustrated by Orit Halpern's work on the history of perceptual machines.[18] There and here, we acknowledge the complexity and the many actors involved in establishing a process. Vision cannot be attributed to a single organ. It requires a complex combination of lenses, nerves, cells, light refraction, and more. Claude Shannon and Warren Weaver's theory of communication is a helpful way to highlight the ways in which light as pure information is communicated to the brain and the role of spectacles in the processing of information. The first part of the mechanism is that a data source generates the message.[19] For the purposes of this explanation we will take the lightbulb as the source of information. The second part, where one or more senders translate the message into a signal, is where spectacles (or the lens) come into play.[20] To further elaborate the role of the lens as a sender, consider Kittler's comparison of the eye to the camera obscura.[21] Generally speaking, the camera obscura functions as a noise filter; it eliminates the omnipresent scattered light and forces only the desirable rays through a small hole.[22] It is able to produce a refined image. This comparison largely illustrates the "naked thesis," wherein Kittler argues that we do not understand our senses until we have media to create models and metaphors for us.[23] Comparing technical media with eyes has a long history. Renaissance artist and engineer Filippo Brunelleschi offered such a comparison, noting that perspective vision always occurs in the eye,[24] although it has been suggested that perspective is itself a culturally specific way of seeing that does not, for instance, take account of those who live in a "circular world."[25] Furthermore, Brunelleschi's work highlights the operational nature of the eye, illustrating that the eye is replicable.[26] Specifically for this chapter, we look at the ways in which spectacles are able to replace, enhance, or correct aspects of the biological optical mechanism. What is essential to note is that the spectacle lens as well as the iris and biological lens function much like a camera obscura, filtering the scattered light, focusing specific rays, and altering their direction so as to produce a clear image on the retina. For individuals who are farsighted, corrective lenses focus light rays to form a focal point on the retina. If the corrective lenses were not present, the focal point would develop beyond the retina, forming an unclear image. The opposite is true for those individuals who are nearsighted. In this case the focal point is before the retina, and the corrective lenses work to alter the light rays so that the focal point is pushed back onto the retina.

The second part of Shannon and Weaver's theory looks at where multiple senders translate or encode the message. The corrective lens is only one of those senders; others include the cornea, iris, and biological lens, which work in similar ways to the corrective lenses, refracting light and focusing specific beams for the production of an image on the retina.[27] The retina itself also plays an essential role in encoding the message, filtering visual information.[28] The retina is a thin membrane at the back of the eye composed of rod and cone cells, nerve cells, and bipolar and horizontal cells, as well as amacrine and ganglion cells.[29] Each cell and nerve functions as a cohesive unit to produce an image of the outer world on which the eye is focusing at that moment. In terms of physics, the photoreceptor cells (rods and cones) absorb the photons (the quanta of visible light) and produce an electrical signal "that allows the encoding of the optical image into a neural image."[30]

Transmission of the encoded message travels through the optic nerve to the visual cortex for decoding. This is the third part of Shannon and Weaver's communication theory, wherein a channel conveys the transmission.[31] Through the optic nerve the encoded message passes through the canal, onward to the optic chiasma, and some information crosses to the opposite side, aiding binocular vision.[32] The optic nerve is also responsible for transmitting the encoded message to different parts of the brain for decoding and storage. These are the final two stages of the theory of communication, respectively.[33] The primary visual cortex is largely responsible for decoding the message, and a number of regions in the brain are responsible for both short-term and long-term memory, including the hippocampus.[34]

We have outlined the optical mechanism in detail here to illustrate the numerous parts that make up the sense of vision in the human body. The role of the corrective lens is therefore a small—filtering data before it reaches the eye—but significant part of the process as a whole. Further encoding and transmission only occurs biologically after the lens's filtering of photons. This process is explicitly highlighted to indicate its essential role in the daily practices, routines, and process of the individual. Similar to the screens discussed in the previous chapter, the analog lens acts like a sieve, allowing some things through but not others, and the digital screen (in technologies like Google Glass) is circuit based and therefore constantly connected to ensure continuous capture of the subject in the broader network. The processing of information before human interpretation and its direct effects on behavior will be further expanded upon in this chapter.

Corrective Lenses: Origins and Early Developments

Before the corrective lens was an object of the face, it was an object of the hand. Evidence from ancient Roman and Greek culture shows that individuals relied on glass bowls filled with water to magnify letters. Seneca the Younger is often cited using such a mechanism to read the many texts in the Roman library.[35] Subsequent magnifying objects of the hand were referred to as "reading stones"; often made of quartz and beryl, these became commonplace around 1000 C.E.[36] These stones were often used by those with presbyopia, the loss of clear nearsighted vision by an aging population. It should be noted that those affected with farsighted vision before aging were referred to as having hyperopia, but the term was not established until well after the development of both reading and distance glasses. Other examples of corrective lenses exist before the popularization of the reading stone, such as the polished emerald that Nero used for watching gladiator fights. Differing sources speculate that this might have been used to see the fight more clearly, to subdue the bright sun and glare, or to use as a reflection so that he could watch those behind him.[37] In Eastern cultures there is evidence of the use of corrective lenses as early as two thousand years ago. However, it has been noted that early spectacles were used much like Nero's, for protection of the eyes.[38] In 1270, Marco Polo reported an encounter with Chinese elders who used reading stones made from "quartz and precious stones as well as tea-colored glass to soothe conjunctivitis."[39] Due to a lack of available sources, this chapter will predominantly focus on the development of corrective lenses in the Western tradition.

Notably, it was the development of a process for converting sand into glass in the ninth century by Abbas Ibn Firnas, an Andalusian polymath, that pushed the widespread development of the reading stone as it came to be known. These roughly hemispherical lenses were placed atop scrolls and other textual materials for a magnified experience.[40] Reading stones were used well into the thirteenth century in many places across Europe. In 1266, Friar Roger Bacon referred to the use of a reading stone for reading his treatise *Opus Majus,* stating that "if anyone examines letters . . . through the medium of crystal or glass . . . if it be shaped like the lesser segment of a sphere, with the convex side towards the eye, he will see the letters far better."[41] However, scholars speculate that it is around this time that the first wearable corrective lenses were created.

In the previous section we touched on the control afforded to the corrective lens in filtering information before it reaches the eye. The mode of filtration seen in the lens is so similar to the lens found within the eyeball "precisely because [the lenses] were developed strategically to override the senses."[42] The glass bowl filled with water, the reading stone, and spectacles[43] were developed in the hope of overriding unclear myopic vision, to establish a new standard of what clear or preferred vision should be. "Standards determine how media reach our senses," writes Kittler, and the standardization of vision with regard to spectacles will have to be evaluated carefully.[44]

The specific development of eyeglass technology resulted from a combination of antagonistic relationships of technology and sense, as well as the discourse network in which this relationship is set, namely, in those practices of labor in which detailed, nearsighted vision is essential. The following section further explains this network. The development of spectacles due to the development of antagonistic media, or media whose material requirements worked against the affordances of the human body, demanded a response and further development of eyeglasses as media in themselves. Such media are found in the practices of labor involving information infrastructure and knowledge production, such as manuscripts, ledgers, legal documents, and other codices. In other words, an emerging need for extended close work required readers and writers to remain connected to a literacy circuit, and corrective lenses served as a kind of network adaptor that would keep knowledge workers from becoming obsolete due to minor visual deficits.

Presbyopic Corrective Lenses

There is debate over the origins of the first pair of eyeglasses, specifically to whom the fame and recognition for the creation of the wearable technology should be attributed. "The Florentines claim the inventor was Salvino Armando degli Armati. The Pisans insist on Alessandro Spina. The Venetians boast of an unknown craftsman of glass or crystal from Murano."[45] Vincent Ilardi, a prominent historian of the Renaissance, notes that the inventor was most likely an "elderly glass worker who, in the process of handling convex shaped glass disks for making leaded windows, discovered by chance that by placing them close to the eyes, he could see objects more clearly on the other side."[46] Ilardi also speculates that the first pair was created around 1268, but there is no substantial evidence. However, the fact remains that to this day there is a window from 1200 to 1300 in which there is no concrete evidence for who the creator was, from whence he came, or when the first wearable corrective lenses were produced.

While the inventor is unknown, there is evidence that Friar Giordano da Pisa (or da Rivalto) was the first disseminator of the invention.[47] He notably shared the news of the eyeglasses during a sermon in 1306. However, the process of creation would have most likely remained a secret had it not been for one of Giordano's contemporaries, Friar Alessandro della Spina, who also knew the inventor and willingly shared the knowledge necessary to make eyeglasses with many.[48] Early versions consisted of two convex glass disks enclosed in metal or bone rims with "handles centrally connected by a tight rivet so as to clamp the nostrils or be held before the eyes."[49]

With the dissemination of the creation process for the technology, many began to produce their own versions. Guilds were established to attempt to control and standardize production and sales. Venice produced the earliest guild regulation for manufacture and commerce, and the Capitolare dell'arte dei cristalleri "repeated a provision previously recorded in 1284, which prohibited members from making objects of clear glass falsified to resemble rock or quartz crystal and extended the prohibition to non-members as well."[50] Production of spectacles was still a viable option to anyone with the materials available to them, but the sale of the product was strictly controlled by the guild.[51] During this early period there were "many people of various trades making or assembling spectacles who were not registered in state fiscal records specifically as spectacle makers."[52] It should be noted that the manufacturing of spectacles was often not a single craftsman's job. Goldsmiths were often the assemblers of the entire product, not just the makers of the frames, and often the frames were not made of metal.[53] As such, assemblers had to purchase prepared bones from artisans, glass banks from workers, and other materials from craftsmen to be able to assemble the final product; their role resembles that of modern opticians.[54] While goldsmiths were prevalent producers of spectacles, based on prices in the late Middle Ages it would appear that those spectacles manufactured by monks were of higher quality and that top-quality spectacles cost "as much as 60 soldi" and "had crystal lenses."[55] These high-quality spectacles often used gold or silver as their frame as opposed to the more inexpensive wood and bone frames used by those with lesser means. The luxury spectacles were often worn by rich bishops, the less expensive frames by friars and artisans, and by the fourteenth century the non-luxury spectacles were neither "scarce nor expensive."[56] Through archival records Ilardi illustrates the sale of hundreds of spectacles across numerous parts of Italy. These sales referred to those spectacles with frames made of bone, wood, and leather. King Henry VIII of England owned fourteen magnifying lenses that were framed in silver or silver-gilt, and one "garnished with gold."[57] Similarly to Nero, Henry VIII also used a green stone, but he used it to read and not watch gladiator fights.[58]

This brief overview of the early development of spectacles indicates the proliferation of the corrective technologies in association with practices that required detailed vision and standardization. Here we return to Kittler's argument of the development of technology due to antagonistic relationships between technologies and the senses and between technologies and other technologies. The development of eyeglasses was a response to both types of antagonistic relationships. The first response we will reexamine relates to the senses. During the early period of their development and well into the popularization of spectacles, they were being used as part of a system of objects relative to those professions who had their "operation center in the hand."[59] These professions included, but were not limited to, copyists, engravers, calligraphers, teachers, merchants, miniaturists, notaries, judges, goldsmiths, spinners, weavers, embroiderers, shoemakers, tailors, scribes, and monks. A noted number of professions dealing with the production of texts such as those found in monasteries and universities, which included activities such as writing, reading, translation, and the production of books, are also well known for their use of the early spectacles.[60] These professions that included practices of working with small details and objects elicited the production of convex lenses well before the development of concave lenses. Tomás Maldonado, a prominent design theorist, argues that it was this relationship of nearsighted labor with a wide array of professions that demanded the production of convex lenses. This developmental relationship is also in large part due to the decay of vision with age. When men reached forty to fifty years of age, their ability to work in these fields was often limited if not lost completely. The use of the convex lens therefore allowed a working class to remain within their professions, and as such it reduced the need to constantly train a new labor force.

The relationship between age, professional practices, and vision is illustrated in Umberto Eco's first novel, *The Name of the Rose.* The protagonist, William of Baskerville, is a man in his later years who relies heavily on his own spectacles throughout the murder mystery. The story, which takes place at a monastery, heavily emphasizes the role of corrective vision, presbyopia, and the standardization of technology as essential to man. The investigation itself is halted when William's spectacles are stolen. His old age inhibits him from reading the names of the books which act as clues, and it is not until he receives a new pair of spectacles that he is able to follow the investigation to the end.[61] William explains to the novice accompanying him that "when a man had passed the middle point of his life, even if his sight had always been excellent, the eye hardened and the pupil became recalcitrant, so that many learned men had virtually died, as far as reading and writing were concerned, after their fiftieth summer."[62] Eco's novel, alongside the works of Maldonado and Ilardi, highlights the close relationship of text, aged vision, monasteries, and spectacles. The emphasis on the use of spectacles for reading by monks and other religious figures plays heavily on their practices and demands. As one of the main sources for the production of texts, as well as centers for knowledge, the practice of reading "silently and alone was developed as early as the seventh century in monasteries and thirteenth century in universities."[63] These practices demanded more of an individual optical mechanism, which came to require the aid of the wearable reading stones. As such, without the technology, readers and scribes were inhibited from their knowledge production and consumption, shaping their daily activities, societal roles, and in some cases, like William, their independence. Later in this chapter we will see similar patterns of dependence and information consumption with the modern digital variation of spectacles in the form of Google Glass and Microsoft HoloLens.

The relationship between reading, spectacles, and religious figures is in part so well defined due to enduring art and literature that note the early relationship between the three. The first known portrait of a person wearing the earliest form of spectacles is that of a French Dominican, Cardinal Hugh of St. Cher (c. 1200–1263).[64] Ilardi argues that it can be assumed that "the diffusion of the invention must have radiated rapidly and widely within the much-traveled community of monks, scholars, and merchants."[65] What Ilardi's point and *The Name of the Rose* have in common is not simply the individuals involved but the practices of these individuals and their subsequent productions. As places of knowledge creation and consumption, monasteries played an important role in the production of manuscripts throughout the Middle Ages before universities took on the responsibility. The relationship between manuscripts and spectacles cannot be overestimated—more specifically, the antagonistic relationship between the two. The development of spectacles can be viewed as a response to the manuscript medium. Between the eleventh and twelfth centuries there was a 263 percent increase in the production of manuscripts across western Europe.[66] In the following century there was a 129 percent increase in the production of manuscripts, with over one million copies produced.[67] We can draw two conclusions from these statistics. First, manuscript production throughout these two centuries occurred without the aid of the printing press. The labor of production was mostly carried out in monasteries by monks and friars by hand. As illustrated previously, the practices of inscription, especially by those in an aging labor class, often required a visual aid for magnification. The increasing rise of manuscript production in the two centuries prior to the creation of spectacles therefore demanded a technological advancement that would allow those working with the inscription media to continue their work. Second, where there is inscription there is reading of said inscription. With well over two million manuscripts produced in two centuries, the knowledge consumption was also increasing. As such, those with literacy skills would also come to require such a technology for a more convenient practice of reading. Eco illustrates the practice by writing that "William preferred to read with these [spectacles] before his eyes, and he said they made his vision better than what nature had endowed him with or than his advanced age, especially as the daylight failed, would permit."[68] It may also be that this need for spectacles by those performing the labor is largely the reason why monks, especially those in Italy, were among the best and most prolific manufacturers of spectacles.[69] As such, spectacles were developed due to their antagonistic relationship with technologies of inscription such as manuscripts. Put differently, the knowledge demands of a particular power relationship necessitated an alteration in technologies of the self.

Further evidence illustrates the relationship between the practice of reading and the dissemination of spectacles into the public at large. The introduction of Johannes Gutenberg's printing press and the subsequent increase in literacy in the public is also closely tied to the proliferation in spectacle manufacturing and sales.[70] Ilardi argues that the Protestant Reformation's emphasis on Bible reading in tandem with the printing press resulted in "the use of eyeglasses in all trades and professions assum[ing] more massive proportions."[71] The religious institutions and culture in Europe throughout this period worked to establish such practices that would require the need of corrective lenses. Kittler highlights a similar relationship, noting that Einstein argued that with the proliferation of copies of drafts and other technological information there was an increase in engineering and other technologies.[72] The relationship between the Gutenberg press, literacy, and spectacles is much the same. The more copies to be read, the more labor to do, the more technologies of assistance are required. As such, the production and proliferation of eyeglasses was in response to the antagonistic relationship between labor and the senses or other technologies.

Myopic Corrective Lenses

The development of spectacles did not end with the corrective reading glasses and their many frames. The frames that developed through trade and practices include, but are not limited to, the first model of spectacles that had two arms, which folded above the nose and had to be held with a hand at all times. These were often referred to as scissor glasses. Through trade, the scissors were altered to include rivets on the side whereby a ribbon would be inserted to tie around the ears.[73] Later a headband was added to the spectacles for free-hand motion, and in the 1600s the pince-nez were developed that sprung close to the bridge of the nose. Subsequent development included the monocle in the 1700s, as well as the lorgnette—a frame with two corrective lenses held up by a stick—also in the 1700s.[74]

Approximately 150 years after the development of reading glasses, corrective lenses for those with nearsightedness were created. There are earlier examples of myopic lenses before those of the 1450s. While different sources provide different reasons, Nero's polished emeralds stand as one of the first traditional pairs of spectacles intended to correct myopia. However, Nero's emerald, as previously stated, may have had other uses as well. The origins of the myopic corrective lenses are as obscure as the presbyopic spectacles. Simone Nerucci, an Italian craftsman who possessed the secrets and tools for making glasses, is credited with finding a more efficient way of "grinding and polishing convex lenses graded to age category and perhaps even concave lenses to correct myopia."[75] While Nerucci is not the originator, it is commonly acknowledged that the first pair of myopic spectacles was also produced in Italy. Scholars questioned what the limitation was that inhibited the creation of corrective myopic lenses until the 1450s.[76] Notably, Maldonado questioned this developmental chronology of spectacles for hyperopia before myopia, drawing connections between practices of labor as well as the technologies of convex and concave lenses. He argued it is the organization of the division of labor that required the response to hyperopia before myopia. Maldonado asserts this development as a response to the aging workforce that would have to migrate to farsighted labors, such as hunting after the age of forty to fifty due to presbyopia. Those who were farsighted had to work in areas where "good long-distance vision was indispensable such as hunter, farmer, shepherd, livestock breeder, fisherman, woodsman, mason, miner, sailor, and soldier."[77] However, Maldonado also points out that throughout the thirteenth and fourteenth century, the practice of dividing the labor force according "to the visual capacities of individuals to see things up close and at a distance was no longer regarded as the most suitable to deal with the changes that were slowly (but inexorably) happening in the society."[78]

Maldonado does acknowledge the technological determinist thesis that the chronological development of differentiated corrective lenses was due to "the simple fact that the craftsmen [and] opticians were not capable, before 1450, of producing concave-diverging lenses," but only in part.[79] The development of corrective glasses for myopia was the "result of a rare temporal coincidence of two factors": the first was the reflections of "Oxonians, Robert Grossatesta and Roger Bacon, on the optical properties of convex lenses," and the second was the construction of similar lenses by the Italians.[80] However, we argue that while the technological determinist thesis holds weight in this argument, the effect of the antagonistic relationships must be considered in order to fully comprehend the chronology of development of spectacles for presbyopia and myopia.

In the development of the presbyopic corrective spectacles there was both the antagonistic relationship between the senses and spectacles, and technology in the form of detailed work and spectacles. However, there was little of this antagonistic relationship for myopic corrective spectacles. It therefore makes sense that the greater of antagonists would be resolved first. Those with myopic vision were "a small minority at that time and they were not totally handicapped."[81] Unlike those with presbyopic vision, who could not magnify the size of their work by moving closer or moving away, those with myopic vision could move closer to the object of their gaze to view it more clearly. These individuals are also well suited for professions and practices that include fine detailed work, such as those crafts mentioned earlier. Individuals were often not afflicted with presbyopic vision until into their forties or fifties. Myopia on the other hand, seemed to afflict a younger population than presbyopia. Ilardi illustrates that "the first dozen glasses ordered were to be fitted with concave lenses for myopic young persons."[82] This order comes from a 1462 request from the Sforza court indicating by the casual wording that concave-lens glasses for young individuals were available in Florence at an earlier date.[83] Corrective lenses were developed, therefore, in response to a moderately less antagonistic relationship to the senses. However, there is little evidence of the technological response to other technologies. Based on documentation, by the end of the fifteenth century, Italy alone generated sufficient documentation to prove the "widespread use of spectacles both for myopes and presbyopes than any other country in Europe."[84] The use of myopic corrective spectacles was not limited to the youth; King Henry VIII used corrective glasses at the age of twenty-three to "see more clearly the rapidly approaching adversary," and Napoleon Bonaparte used handheld scissor spectacles to "combat his shortsightedness."[85]

The combination of concave and convex lenses in a single pair of spectacles did not emerge until the late 1700s. The idea of the split lens was suggested by Johann Zahn in 1683, and once more by Christian Gottlieb Hertel in 1716. However, there is no evidence of such lenses being used in a practical ophthalmic function until decades later.[86] Benjamin Franklin has long been popularly credited for the invention of the bifocal. There is insufficient scholarly evidence to prove this claim. Further, during Franklin's diplomatic mission to France he ordered a pair of bifocals from the English optician Sykes, who wrote to Franklin apologizing for the delay due to their breaking three times during the cutting process.[87] This has led to speculation that bifocal glasses were not a commonplace order for the time.[88] It is more likely that a number of individuals were working on such lenses simultaneously, based on ideas brought up by Zhan and Hertel. The term *bifocal* was not used until John Isaac Hawkins, the inventor of trifocals, came up with the term. Trifocals were created in 1827 to suit the needs of those with myopia and presbyopia; the third distance was an intermediate distance—approximately arm's length—to aid aging eyes.[89]

In his seminal book *Understanding Media,* McLuhan argues that, among other effects, "TV makes for myopia."[90] By the 1960s television already had a strong presence in the lives of many westerners. However, it was the development of computer screens that caused a new set of antagonistic relationships that incited the further development of eyeglasses as a technology. LED (light-emitting diode) screens emit blue light (400–500 nm) that is harmful to the human eye.[91] The average Western individual can spend close to twenty-fours a week engaging with a digital screen, and therefore absorbing a good deal of blue light.[92] As such computer glasses were developed to respond to this harmful practice, most notably in part due to the large workforce that has come to depend on digital practices. Computer glasses, often with yellow-tinted lenses, reflect blue light to offer relief to those individuals suffering from digital eye strain (DES). These computer glasses function similarly to the protective glasses patented by the Swiss Accident Insurance Institute in 1929 in response to the one out of seven safety incidents that concerned the eye.[93] However, recent research indicates that the blue light emitted from digital screens is of low dosage and that even across a long period of time it is not a biohazard.[94] Furthermore, the effectiveness of computer glasses to eliminate the symptoms of DES has not been proven, and further research is currently being carried out.[95]

Smart Glasses

We do not claim that there were no developments in spectacles since the Renaissance. The twentieth and twenty-first centuries saw the production and development of spectacles with reality-augmenting and virtual-reality capabilities in response to practices of labor and other technology. The first representation of reality-augmenting technology is a pair of spectacles the Demon of Electricity gifts to a young boy named Rob in L. Frank Baum's *The Master Key: An Electrical Fairy Tale* in the 1901 novel. The spectacles are referred to as the Character Marker, which while worn "every one you meet will be marked upon the forehead with a letter indicating his or her character. The good will bear the letter 'G,' the evil the letter 'E.' The wise will be marked with a 'W' and the foolish with an 'F.' The kind will show a 'K' upon their foreheads and the cruel a letter 'C.' "[96] How the spectacles are able to achieve such a feat and where they access their information is not explained, leaving the spectacles as a hybrid object of electricity and magic. What is perhaps most interesting about the novel is the ending: Rob tells the Demon that humanity is not ready for the electrical gifts and that the Demon should keep such objects until a time when humans know how to use them.

That time arrived some fifty years later. Augmented-reality (AR) and virtual-reality (VR) technologies were developed throughout the twentieth century. In 1962 a cinematographer named Morton Heilig patented the Sensorama. It was a simulator with visuals, sound, vibration, and smell[97] that produced a more immersive cinematic experience. In 1966, supported by the Advanced Research Project Agency (ARPA), Dr. Ivan Sutherland invented a three-dimensional head-mounted display (HMD).[98] In the 1980s, Dr. Steve Mann began developing wearable AR technology that would become the basis of the EyeTap, a head-mounted technology that simultaneously records all that the eye sees and superimposes information on reality. Mann's work is more closely related to the reality-augmenting spectacles used in our contemporary world. In 1992, at the U.S. Air Force Research Laboratory (Armstrong), Louis Rosenberg developed one of the first functioning AR systems. He called it Virtual Fixtures, a robotic exoskeleton that the user would wear to improve human performance. What follows is a plethora of research and development on AR and VR by academic institutions, military forces, and the entertainment industry. It is not until May 15, 2014, that the general public gains access to Google Glass, a pair of smart spectacles like those described in *The Master Key*, and this time the public knows what to do with them, or at least we're figuring it out.

Smart spectacles—or smart glasses, as they are more often called—are wearable computer glasses that have the ability to augment reality by adding information to (or on top of) the wearer's field of view. The difference between analog and digital eyeglass technology lies in the technology's ability to select, store, and process the information of any given discourse network. Although analog spectacles have the ability to collect light as information and simultaneously process the light rays before they reach the human eye, they lack the ability to store information. Without the ability to store information for later recall, regular spectacles' ability to process information is limited to noise cancellation. Their collection of information is also limited to the physical rays of light in the proximity of the lenses. With smart glasses, the shift in the three abilities of optical technology becomes profound. This is mostly due to the smart glasses' connectivity to the internet, which allows them to access information not only in the physical world but also information found on the web. Smart spectacles integrate digital knowledge into the physically recorded world. Collection from the internet, storage linked to a cloud, and the processing of information no longer come down solely to the lens, but to a combination of hardware and software.

Smart glasses often run self-contained mobile applications for a number of practices, including but not limited to fitness, security, entertainment, and labor. As HMDs, smart glasses are often hands-free, voice activated, and include small touch buttons, or a combination of the three. James Melzer and Kirk Moffitt argue in *Head-Mounted Displays: Designing for the User* that an HMD "can be personal, interactive, expansive, and virtual."[99] It is the personal nature of the HMD that has allowed certain discourse networks pertaining to functions of collection, storage, processing, and productivity to come about. However, this development from AR technology to wearable computing was not the result of a string of coincidences. Rather, the development of smart glasses, like that of analog spectacles, comes down to the antagonistic relationships found between technologies and between technology and the senses.[100]

The first case is the development of smart glasses due to an antagonistic response to technology. There is no one technology to which we can attribute the antagonistic relationship. A combination of cinema, war efforts, and academic pursuits contributed to the development of AR technologies and digital spectacles. However, if we were to name a single medium as the antagonist that spurred the production of smart glasses, it would be the internet. Long gone are the days of dial-up; the smartphone itself is now required to compete with wearable technology, which requires no tether, is instant, and is always in the field of view and never in the back pocket.

This is an echo of the same process analog glasses took part in throughout their development. The object of the hand, in this case the laptop, tablet, or cell phone, is beginning to see competition from an object of the face, the smart glasses. Isabel Pederson sums up the rhetoric often associated with head-mounted displays and virtual environments as "humanizing utopian, because they were first conceived to solve many problems associated with desktop computers."[101] In media's constant struggle to one-up each other, humans are relegated to the role of the means of production. Smart glasses are merely the latest response in a long line of antagonistic medial relationships.

The second case is the development of smart glasses due to an antagonistic response to the body. If there was ever an example of technological development that was strategically developed to override the senses, as Kittler argues most technologies are, it is reality-augmenting glasses.[102] Smart glasses aim to override, manage, and otherwise negotiate the human sense of vision. Perhaps the most obvious method of doing so is their ability to overlay the field of view with information not found in physical reality, but a better way to explain it would be to follow Shannon and Weaver's model of communication and the examples of the camera obscura to better understand the process of how the Glass technology works.

Referring back to our discussion of the analog spectacles, readers may recall that the lenses function in the same way as the camera obscura by acting as a noise filter for light rays. Although Glass has the ability to reach a near-infinite amount of information, it acts as a noise filter to offer only the desirable information for any given user. Who decides what information is desirable is a different question entirely. Here we refer back to Brunelleschi's work, which illustrated the ways in which the eye was replicable. Glass functions as an eye before the eye and a mind before the brain. In the analog model, the corrective lens acts as one of the many senders that encode the message in the second step of our communication model. In the smart glasses version of the model, the lens functions as all five steps of the model (data source, senders encoding the message, transmission of the encoded message, decoding of the message, and memory). However, due to Glass's proximity to the eye and its reality-augmenting nature, the image on Glass is simultaneously paused at the third state of transmitting the encoded message to the biological eye, where further processing, decoding, and memory are achieved. This duality has been the cause for critical inspection and in some cases public rejection for fear of manipulation and control. Its function as a technique/technology of control and of the self is a topic we will be interrogating throughout the remainder of the chapter.

Information Processing and Discourse of the Fallible Brain

Technologies of control are specific to their discourse networks. To effectively explain how reality-augmenting glasses are technologies of control and of the self, we will present three case studies. Each case study will attend to one of the three components of a discourse network: selecting, storing, and processing. If smart glasses were developed to override the sense of vision and visual processing, it is because the biological process is seen as a problem that requires optimization. Western culture, now driven by the knowledge-making apparatus that privileges big data, demands a populace to process, store, and collect more information at any given time. Google Glass as a technology is used to both perpetuate the discourse network of the cultural value of information and more data, as well as acting as the means through which humans are able to select, store, and process greater quantities of information. Wearable technology has become an integral part of the daily life of the digital human. Here we examine how smart glasses, including but not limited to Google Glass, HoloLens, and others, shape and control the scale and form of human behavior and action. If the medium is the message, then smart glasses are a mass information control center. Smart glasses control the scale of information that becomes accessible to the wearer and shift human interactions with the internet to a more immediate and constant relationship. Wearable technology is therefore often portrayed as the means to control an overabundancy where the biological nervous system cannot. Yet these technologies do not offer an expansion of human collection, storage, and processing capabilities; rather, they expand the information system, decentering the human to the subject of information production.

Consider Jorge Luis Borges's tale "Funes, the Memorious." Borges relates his encounter with Funes, a man able to remember every detail of raw data he ever encountered. Funes tells Borges that he had lived like a person in a dream: "he looked without seeing, heard without hearing, forgot everything—almost everything."[103] Essentially, Funes had lived like the rest of us, taking in a limited amount of information, storing some of it, and only able to process a fraction. After falling from a horse and receiving an injury, Funes gains the ability to see, hear, and remember everything.

He becomes an organic supercomputer and states, "I have more memories in myself alone than all men have had since the world was a world."[104] Funes's issue arises not from his ability to collect and store so much information within himself, but to his ability to process it. Borges illustrates this when he explains that Funes's goal is to "reduce all of his past experience to some seventy thousand recollections, which he would later define numerically."[105] When there is too much content, Funes filters it into something more manageable—succinct quantified data. Borges's portrayal of Funes is one common to the discourse of HMDs: the optimized human brain would have the ability to collect more information, pull at stores of memory, and process it in a timely manner.

Reasoning for the optimization of the practices of collection, storage, and processing of the human brain belongs to a history of cybernetics and psychology. In the 1950s, with the development of cybernetics programs and the expansion of psychological research, logic and reasoning became comparable to computers and algorithms. Orit Halpern argues that with the cybernetic reformulation of the social and behavioral sciences the mind was explained to work like "nets, with series of steps leading to circuits, then these circuits could be discretized, modeled, rebuilt, reenacted."[106] Halpern goes on to point out that in "cognitive science, as in neuroscience, cognition and perception were rendered equivalent—both treatable as communication channels, and subsequently both subject to new forms of intervention."[107] As such, vision becomes an "algorithmic process" that can be manipulated by reality-augmenting input to establish new circuits of cognition and perception. (Governing a body through the manipulation of the "rebuilt" mind is examined further in the following "Processing" section below.) What is imperative to recognize about this work is that although it focuses on research and application done throughout the twentieth century, the discourse of vision and "broader changes in governmentality relating to how perception, cognition, and power were organized" are still influential today in relation to smart glasses and information processing.[108]

The expectation is that more information allows the individual to predict and prevent future undesired situations. The more information we have, the more we can control what the outcome will be, but as Richard Lanham illustrates, "information doesn't seem in short supply. Precisely the opposite. We're drowning in it."[109] Yet information systems are constantly attempting to expand. If the biological system can collect, store, and process only so much, and if it cannot add all the information it manages to a system, then technologies will shape humans to better aid the information system in proliferating further. Wearable technology brings us in only to make us insufficient to the stores of data running through the circuit. With analog spectacles, people were shaped by the media to perform tasks such as reading, writing, and other small detailed work well past the age where biologically they could. Smart glasses mold the human to depend on the wearable technology by simultaneously producing and consuming more information while increasing mobility.

Inspired by Borges, Mark Andrejevic notes that "every day we are bombarded with more information than we can possibly absorb or recall."[110] In our attempts to absorb the onslaught of information, we tentatively walk the line separating medial madness from medial control. Andrejevic defines the culture that is both responsible for the creation of the smart glasses and bolstered by it as a digital enclosure. Essentially it is "the creation of an interactive realm wherein every action, interaction, and transaction generates information about itself."[111] Just as Funes, who recalls every time he remembers a day and adds that recall to the memory, so that every time he recalls the memory he also remembers every other time he has recalled the memory, it becomes cyclical and exponential in the production of information. Smart-glasses-wearing humans help their digital enclosure grow exponentially through this same practice. In turn, this makes humans further dependent on the technology that offers such information consumption and management, as will be touched on in a later section of this chapter.

Google Glass

In May 14, 2013, Google began selling the prototype of Glass to those companies that qualified as Glass Explorers. Glass is an optical HMD designed to look like a pair of analog spectacles without the corrective lenses, but with an additional prism. It was developed by Google X (now X) in an effort to produce ubiquitous computing.[112] Like most ubiquitous computing dreams, Glass is designed to be personally interactive and respond to natural voice commands, the equivalent of "Look ma, no hands!" This technology was to be adapted to the Google Home technology, a voice-activated speaker that offers the ubiquitous computing home described by Mark Weiser in "The Computer for the 21st Century."[113] Weiser presents his ideal of ubiquitous computing by offering readers a ride-along experience with Sal, who wakes, commutes, and works in a fully ubiquitous computing world, never once being isolated from the internet of things that makes up all the technology surrounding her. Not surprisingly, ubiquitous computing is desirable for its accommodating nature. In his explanation of ubiquitous computing, Weiser notes that for information systems to be used by humans to even greater extents, the technology must disappear from our consciousness. He refers to Martin Heidegger's "horizon" and "ready-at-hand" as well as Herbert A. Simon's phenomenon of "compiling" to explain this, but the argument is simple: the medium must be invisible for the user to see beyond it to other goals.[114] When technology becomes invisible, it is easier for people to use, and ubiquitous computing reaches its pinnacle. Google Home offers ubiquitous computing only in the home. Google Glass offers the mobility necessary to make the whole world connected, as long as there's WiFi available.

Glass is not ubiquitous simply because it is connected to the internet. Rather, the applications developed by Glass Partners (previously "Glass Explorers") make Glass a ubiquitous experience. Early models of Glass indicated that it could be used by the average person to do things such as taking pictures and videos, texting and calling, and pulling up geographical information. But this is not where it ends. Glass Partners have used Glass as a platform to augment those three factors of discourse networks: selecting, storing, and processing information in relation to optical mechanisms. Finally, Glass is being used to optimize the human using all three factors in professional settings. If analog spectacles were referred to as corrective lenses, the rhetoric of smart spectacles is abundant with possibilities for correction and improvement of behaviors and tasks because of an extra layer of surveillance, capture, and post-processing. It is the optical nature of Glass, in its practice of making the human the object of information through its geospatial configuration of viewing, and not simply the subject of communication, that marks it a technology of the self. How are we using collection, storage, and processing as a technology of self, to care and manage the self through Glass? Perhaps it is most apparent when considered in its specific discourse networks. We will consider three case studies to better explain the relationship between technology and the human, focusing on applications that have been used to aid humans in collecting visual information, storing and accessing information, and finally processing it through Glass. Medial madness can only occur through the selection, storage, and processing of information. When media intercept our preexisting circuits to direct and modulate new flows of knowledge through these information processes, they flood previous flows to determine what constitutes the new reality. Smart glasses intercept the visual and memory circuits, overriding our meaning making practices to establish new meaning, leaving the subject vulnerable to a state of medial madness.

Selecting

In *Gramophone, Film, Typewriter* Kittler posits that the "eyes have become autonomous" with the production of storage technologies that are able to record and reproduce the time flow of optical data.[115] This is certainly the case when considering the selection of optical media in a discourse network. In previous chapters, and in this chapter's discussion of analog spectacles, we have referred to the three characteristics that make up a discourse network. Selection of data is an essential step to the production of information, culture, and the management of the self.

In *The Spectacles,* Edgar Allan Poe comically illustrates the effects of the limited selection of visual data when a twenty-one-year-old man mistakes an eighty-two-year-old woman for a young girl and attempts to marry her.[116] The young man, who refuses to wear visual aids throughout the tale, is given information by those around him of the reality in front of him so that he is constantly using both audio and visual information to create an image that is not a correct representation of reality. What this tale not-so-comically highlights, however, is how the selection of data can be manipulated to produce an image that does not exist for those with visual impairment. Selection of optical data—the intake of light photons—is what establishes our reality. When we depend on technologies and their corresponding circuits to select our optical data and create our reality, our behaviors are then molded and shaped by the very circuits we depend upon.

Like the history of analog spectacles, the media history of smart eyeglasses is a story of media convergence. Technologies of visual assistance are not limited to spectacles. The rich history includes an early edition of the typewriter, created in the early nineteenth century by Pellegrino Turri for Countess Carolina Fantoni da Fivizzano (1781–1841), whose vision was slowly fading, so that she would be able to correspond with her friends, but more importantly with him privately. "When the Countess pushed a key, an arm struck a piece of carbon paper atop a sheet of paper."[117] In 1872,

Thomas Edison patented an electric typewriter, and in the 1920s the first workable model was introduced. Since the 1930s, the development of the typewriter became intertwined with the development of the computer, as the typewriter was the only mechanism for inputting data.[118] Other technologies of visual assistance relied not on touch but on sound. In 1962 IBM released the Shoebox, a technological medium capable of recording and processing sixteen words and the digits zero to nine. Thirty years later, DragonDictate released their own program for speech recognition. These assistive devices were also developed in an effort to offer greater accessibility to those with visual impairments. "Early observers often likened the Internet to assistive devices for the visually and audibly impaired."[119] A combination of spectacles, touch technology, and voice recognition have been developed to offer a more powerful assistive technology that allows those with visual impairment to experience the world around them in a different way.

If we consider that the eyes are the only organ physically capable of processing light to produce visual information, and that Glass is now able to do it for those with limited vision, then the eyes have certainly become replaceable. One Glass Partner, Aira, has developed a system that aids the visually impaired to "see" their surroundings. Users connect through an application on their smartphone (touch), that connects their Google Glass (spectacles) to an agent who, through the camera on Glass, is able to see everything around the user and offer real-time description (speech recognition) to create an image for the user of their reality. Aira's mission is to provide users with "on-demand access to visual information."[120] Like many other Google developers, Aira supports efficiency, engagement, and independence in the user. These non-obtrusive spectacles "offer a minimal material demand on a human user, but offer a wealth of experience reaching far beyond their seeming abilities."[121] It is essential to recognize that while this is a technology of control and of the self, it does offer important, beneficial support to those who need it. This circuit of phone–agent–Glass certainly offers a wealth of experience, but it also brings to mind questions of governance. As explained in the Introduction, circuits enact differing modes of governance, whereby they guide conduct. Individuals' behavior is affected by their environment, the social codes of conduct set for the environment, and the knowledge that someone else is seeing what they do as they do it. Aira's agents, in conjunction with those apparatuses in the circuit, have complete control of the reality-building experience and can create a reality that does not exist, or even a simply modified reality for users, thereby conducting their conduct. Glass and the circuit are overriding the visual-cognitive process to offer a precognition, in which they produce a vision of the world as they see it.

Storing

"If to convert to oneself is to turn away from the preoccupations of the external world, from the concerns of ambition, from fear of the future, then one can turn back to one's own past, recall it to mind, have it unfold as one pleases before one's own eyes, and have a relationship with it that nothing can disturb."[122] For some, however, turning back to one's own past is no longer possible. Whether through injury, illness, age, or simply forgetful moments in passing, the human memory is not as consistent as we would like it to be. Technologies and techniques of memory augmentation have been in use since the classical period when Greeks and Romans used memory palaces as mnemonic tools to memorize large amounts of information for rhetorical treatises. Google Glass is not the first technology to store sensory data. Kittler points out that even pre-internet media such as silent film had the ability to store visual data.[123] Glass is merely the latest in a long procession of memory-aiding media. From sticky notes to calendars to journals and pictures, visual memory aids have become a commonplace object in the lives of many. As information systems proliferate, media respond to create new memory aids. This has become especially important for those with Alzheimer's, traumatic brain injury (TBI), or any variety of health afflictions affecting memory.

Readers may question the validity of a technology capable of storing memories. Pictures and film store visual data, but they do not hold the contextual metadata in addition to the visual recording. This is the case for individuals who lose a memory and look at a picture to help recall it but do not know who is standing next to them in the image. Analog media retained the capacity to store some sensory data, but not enough to constitute the human memory. Kittler argues that "electricity itself put an end to [the hallucinations of writing]" and made memories technically reproducible.[124] And while we have the technical ability to do this, to reproduce a whole life through technical means would be a vast undertaking. Here we face the same problem as Borges's Funes. The storage of past accounts is not sufficient if the subject cannot process the information or use it again.

Funes processed his vast memories by using numbers. In the digital age, we have learned to do the same with the aid of only two numbers: zeroes and ones.

Memory can be divided into two functions. The first is storing data; the second is operating on that stored data.[125] Kittler argues that it was electricity that allowed us to reproduce memories, but it is digital innovations that have allowed us to operate those memories. According to Halpern, the work of George Armitage Miller, one of the founders of cognitive science, "opened the path to the augmentation, and perhaps automation, of both memory and decision-making."[126] Cognitive science, under the influence of Miller's work, turned its attention to the channel capacity of human perception, and its "primary concern was enhancing the subject's ability to consume information."[127] When it comes to visual processing, memory plays a part in steps four and five of Shannon and Weaver's communication model. Certainly, it is the storing of information, but it is also necessary for contextualizing and therefore decoding information. With the development of Google Glass and corresponding glassware, among other technologies, we now have the ability to intercept in the final two steps of visual processing, whereas before it only took part in the first three.

Morgan Barense and her team have been creating a virtual hippocampus. Barense is the Canadian Research Chair of Cognitive Neuroscience, and an associate professor at the University of Toronto. Her research has culminated in a phone application called HippoCamera, that replicates the role of the hippocampus by recording memories through video, and playing them at high-speed with the dictated oral description of the subject the way the hippocampus would naturally relay memories to the cortex. As a device the HippoCamera falls somewhere between Glass and neural implants, and it is currently in testing to help those with Alzheimer's. The death of the neural cells generally starts in the hippocampus, and Barense and her team worked to create a technological prosthesis that would allow the unaffected areas of the brain to continue their memory-making and memory-processing abilities. One cause for concern, according to Dr. Jesse Rissman, a member of the HippoCamera application team, is that "each time we retrieve a memory, it also becomes vulnerable to change."[128] More alarming, he states that "eventually, you might strongly believe that something that didn't happen to you actually did."[129] Science fiction introduces us to technology and its implications long before they manifest, and this concern has been witnessed a plethora of times in works such as *Do Androids Dream of Electric Sheep?* HippoCamera is certainly an aid, but media technologies, as we saw in the previous section, are not always used for the primary intentions of the inventor. HippoCamera is also not the only digital memory aid, and where there are many ways of augmenting memory, there are also many ways of altering them. This system is effective in producing and strengthening memories in that the user records the context of the image or video, thereby controlling the memory process even if it is mediated by Glass and the program.

In 2015, at the International Conference of Artificial Intelligence, Thomas Way, Adam Bemiller, Raghavender Mysari, and Corinne Reimers presented the Electronic Localization, Elucidation and PHotographic Assistive Notification Technology system (ELEPHANT). This system was also developed to help those with Alzheimer's, TBI, and other afflictions causing memory loss. The assistive memory system is based on five components: "The Recollection Triggers module manages user interactions and queries, a Classification Filter manages the memory database, the Memory Item Database where memories are stored, a Recall Evaluator that assesses query results and guides the user to further action as needed, and the Prism Display which includes all Google Glass display, I/O and networking support."[130] These components work in tandem to produce the memory processing found in our biological system. As Pederson illustrates in her analysis of wearable technology, "using body-worn devices, wearable computers, and other technology, a person records information about everything, including every conversation, every body temperature change, every television show watched and every trip taken to every place."[131] The aim of ELEPHANT is not only to assist those with memory loss, like HippoCamera, but also to surrogate the complete process. Way et al. argue that Glass offers a unique platform due to its ability to "easily capture and deliver a wide range of relevant information (visual, audible, and textual) about important subjects in a relatively discreet manner."[132] This discreet manner is helpful to the user throughout their daily practices as they process the information surrounding them. What prompts attention to this application is the "Browse Memories" option in the MemoryScrollActivity. Similar to recollecting memories of days gone by or watching home videos and adding commentary, Browse Memories allows users to access all adjoining information about each of the memories from the database. Interestingly, "if no memories are found, a card is displayed alerting the user that memories should be created."[133] This is where the process diverges from biological memory processing. In fact, this process echoes that of a Boolean query in computer systems: Sorry, match not found.

Such systems are indicative of historical work by Miller and others on cognitive sciences, as well as contemporary rhetoric of need, which is made up of three parts: "(1) memory as an inept faculty, (2) memory in need of surrogacy, and (3) memory as a storage device."[134] Such rhetoric is also indicative of a vulnerability for the colonization of the human memory, the ability to extract further value from human experience in the form of supplementary storage as an extension of the human. Such positioning establishes a reliance on memory aids as prosthetic technologies that define and perpetuate ideas of human limits.

This rhetoric, HippoCamera, and ELEPHANT are a clear indication that the discourse established by Miller in the 1950s is prevalent in computer science and engineering labs today. When the relationship between brain and computer is so closely associated through models of communication, it is not difficult to understand how technology has become the digital "custodian for the human body, knowing what is good for us."[135]

In the same year in Atlanta, Georgia, in the Rehabilitation Engineering Research Center for Wireless Technologies, Shepherd Center, a different group of social scientists were developing a different Glass application to assist memory targeted at those recovering and living with TBI. This system, called EyeRemember, is similar to ELEPHANT in having information readily available for the user thanks to the wearable nature of Glass, but differs from the two previous examples in that it is targeted at remembering people and not memories. This memory aid allows users to "voice-record important information related to individuals in their circle of family, friends and caregivers," and this information is presented when said contacts are in the vicinity or when users scroll through a "timeline" of their information.[136] This system, however, is not as personal as ELEPHANT or HippoCamera; contacts can wear a beacon, and when they are within an established proximity to the user, Glass will notify the user through a visual and audio message that this contact is nearby.[137] This system focuses on helping those with TBI to continue their social interactions with limited inhibitions by offering the information to contextualize their relationship with the contacts. This system is indicative of the practices of self-care established by Foucault in *The History of Sexuality, Volume 3* where he establishes that self-care is "not an exercise in solitude, but a true social practice."[138] Certainly, Glass is a personal technology, but when it comes to memory, it is the structure of relationships, institutions, and other technologies that Glass is meant to relate to the user.

This is not to say that digital memory has only been proposed as a solution for those with mental afflictions; in fact, a plethora of sources both fiction and nonfiction replicate the rhetoric of need for digital memory. In 2003, DARPA had the goal of developing LifeLog, a wearable hardware that would record visual and oral data, and possibly what the user feels through a haptic addition.[139] The project was ultimately not developed, but it is just one example of technologies being created to attend to the "faulty" memory of the human. These examples indicate a growing reliance on digital memory augmentation technologies.

Processing

The collection and storage of information is essential to building a repository of those data points that make up the discourse network. However, without the ability to process that information, the repository becomes useless to the production and reproduction of cultural norms. Of the three aspects, the processing of information, because of its use of the data collected and stored, plays an essential role in the construction of reality and the alteration of behavior. In the optical mechanism the biological circuit manages the visual information to produce an argument: if X image is contextualized by Y data stored, then Z behavior, and in such a manner the processing of visual information is used to govern the self. Foucault's examination of Bentham's panopticon is a modular example of visual data processing for the governance of the self. In the panopticon, the periphery building, which is divided into singular cells, circles the center tower. Each cell has two windows—one that faces the center tower and the other on the opposite wall—so that the light produced by the tower is able to flood the cell as it passes through, proving, as Foucault argues, that "visibility is a trap."[140] The supervisor who stands behind the light produced by the tower is never seen, but able to see all. Light becomes a binary signal for the cell's inhabitant. If the light that pours through their cell allows a supervisor to see all his action, and it is contextualized by the information that the supervisor is able to watch without being seen, then the inhabitant will behave as though he is always being watched. The cell's inhabitant in this sense processes visual data—light on—to govern his behavior based on the stored information previously collected on surveillance. However, not everyone processes visual information in the same way. The merging of their two circuits does not produce the desired response the supervisor expects. The binary still exists, but the second circuit, which often overrides what is in front of us, as previously explained, does not contextualize "light on" to surveillance. Foucault explains:

> For it is not a question of linking consequences, but of grouping and isolating, of analyzing, of matching and pigeon-holing concrete contents; there is nothing more tentative, nothing more empirical (superficially, at least) than the process of establishing an order among things; nothing that demands a sharper eye. . . . And yet an eye not consciously prepared might well group together certain similar figures and distinguish between others on the basis of such and such a difference: in fact, there is no similitude and no distinction, even for the wholly untrained perception, that is not the result of a precise operation and of the application of a preliminary criterion.[141]

Foucault's explanation of the use of observation, a preliminary criterion, and the precise operation of contextualizing and processing the visual information to properly establish order among things also highlights the ways in which things can be ordered differently due to understanding. The inhabitant of the cell may not process the binary light system in the normalized expectation because his preliminary criterion—that is, the stored information that contextualizes optical data—either lacks definition or is not properly applied. As such, those who process visual information differently cannot be governed by the same circuit and its application. One such group that is not governed by the social norms enacted by the circuit are those with autism spectrum disorder (ASD). The *DSM-5* defines ASD as "characterized by persistent deficits in social communication and social interaction across multiple contexts, including deficits in social reciprocity, nonverbal communicative behaviors used for social interaction, and skills in developing, maintaining, and understanding relationships."[142] In their examination of ASD and its relationship to new media, Amit Pinchevski and John Durham Peters argue that "whereas decades of thought about mass media posited face-to-face conversation as the communicative ideal, autism presents an alternate mode of texting, typing, and mediated talk stripped of both verbal and nonverbal complexities."[143] What the classification of ASD and Pinchevski and Peters's work highlights is the non-normative processing of information—optical in this case—and as such the behavior that does not align with the communication "ideals" that govern the individual. As Pinchevski and Peters illustrate, "pathology not only reveals normality, as the doctors have always said, but it also reveals technology."[144]

Not surprising, the latest technology in the history of ASD and media is smart glasses. At the Seventh Annual World Focus on Autism Conference, which took place on September 25, 2015, Google Glass was the focus of the keynote address. The keynote claimed that Glass could get those with autism "to learn to be self-sufficient" or sufficient via technology. One Glass Partner who pursued this venture is Brain Power, an application that works on "behavior control" through "social literacy." The company was founded by Ned T. Sahin, who completed his PhD in neuroscience at Harvard, along with his team of MDs, PhDs, engineers, and therapists. Brain Power is a "neuroscience-based assisted-reality tool" that uses Glass to monitor autistic subjects through small-movement assessment that correlates to various states of emotion and perceptual foci.[145] Together, Brain Power and Glass create a system of information pre-processing for the user. They do this through the camera, microphone, and accelerometer that track head movement. Simultaneously, Brain Power produces images on the lenses so that the user can understand facial expression and certain life "social cues" through gamification.[146]

Referring back to Foucault's argument on the operation of contextualizing and processing the visual information, we can further expand on the role of visual interpretation for the ordering of things. Knowledge processing—that is, the selection and storage of information that constantly overrides our reality—requires a set amount of training to establish the relationship between the two circuits. We are trained from an early age to associate certain facial expressions to certain behaviors. This visual processing becomes a normative structure of governance. Those who are unable to process information in the same way or are predisposed to process visual data differently become ungovernable. Halpern notes that "reality which impinges on a person will exert pressures in the direction of bringing the appropriate cognitive elements into correspondence with that reality," but when reality does not exert pressure on the socially deemed appropriate cognitive response, individuals will act according to their own reality.[147] Brain Power has not released any content or news since 2015, but it is not the only study approaching ASD as a media problematization.

The Autism Glass Project at Stanford Medicine conducted a clinical trial from November 1, 2016, to April 11, 2018, wherein children six to twelve years of age were asked to wear Superpower Glass four times a day for twenty minutes at home in an effort to improve facial recognition and social outcomes.[148] Superpower Glass uses the gamification of facial expressions, emotion recognition, and real-time social cues like Brain Power, but has published the results of their clinical trial, stating that of the seventy-one children with ASD who were treated at home with the wearable intervention, those who received the intervention showed significant improvements on the Vineland Adaptive Behavior Scale socialization subscale compared to those in the control group. The intention of this clinical trial was to determine the efficacy of Superpower Glass in augmenting standards of care therapy to "achieve higher socialization in children with autism spectrum disorder."[149] This is a case of "virtual reality as the bombardment of the senses" overriding our own processing.[150] With the proven efficacy of this circuit in overriding the biological circuit of children with ASD to produce social norms of behavior, questions of re-seeing and pre-seeing the world through Glass become a principal concern.

As established earlier on in this book, Foucault's argument for the rise of a disciplinary society by the prison of the soul and social norms is being pushed aside by a new, totalizing, digital prison. If we will not be governed any longer by the society, the digital will ensure we see the world in a way that enforces expected behaviors and social norms. This is further enacted in the labor industries which govern bodies by pre-processing big data to produce a vision of the world designed by those in control. When we begin to rely on digital circuits to train our basic understanding of the world around us, we relegate our visual information-processing capabilities to the governing systems that control. If reality is being reprocessed and preprocessed how can we trust our vision of the world?

Productivity

If Glass can optimize the collection, storage, and processing of information in discourse networks ranging from injury to illness to development, then it can be used to optimize labor processes. Analog spectacles focused on the labor of reproduction: producing texts, detailed materials such as clocks, and other record keeping. Smart glasses follow the same tradition, being positioned to help professions with small details such as surgery, or production and manufacturing in factory and industrial settings. In addition, the justice system has begun using smart glasses to record live events, as well as facial recognition software.

Glass has been used in a number of professions with the goal of optimizing the preexisting system. Education and labor within the medical field have been one of the leaders in this innovative streak. Surgery specifically has benefited from Glass and its optic-enhancing capabilities, offering students a closer look at a senior surgeon's work in the operating rooms and offering the senior surgeon a closer look at the patient on the table.[151] When it comes to surgery, it is not difficult to comprehend the use of smart glasses. An improved education and training system is not something one would argue against—when the lives of patients are on the line, no one would refute the need for a better look.

China, the country with the "world's biggest camera surveillance network," has added smart sunglasses to their network as of February 2018.[152] The smart glasses have facial recognition software, which China is known for developing, helping officers scan crowds for suspects. Once a suspect is located, the glasses show the officer personal information about the suspect, including birth date and residence.[153] In the short time this technology has been used, it has helped police officers arrest seven suspects of crimes ranging from hit-and-run to kidnapping and human trafficking.[154] Wearable technology has been used before by the Chinese law enforcement to record operations.[155] With an increase in juridical productivity, there is also an increase in surveillance culture. Big Brother is always watching when there are smart glasses involved.

On the factory floor, the harvesting fields, the operating room, or the theater there is a pair of smart glasses available. Microsoft released HoloLens, a mixed-reality heads-up display (HUD), to developers in March 2016. The HoloLens offers—not surprisingly—a holographic experience that overlays the physical world. Microsoft has been working on this technology for a number of years and an early model of it was seen with Xbox Kinect, a successful gaming console and application. The HoloLens is in its second development and has been sold across the globe. The HUD works through gesture and voice command. To open and close a screen, users can use a single hand. They can move from one screen to the next to a real-world object using only their hands. The HUD can also be voice activated, and works with the gaze of the individual to create the augmented experience. What distinguishes Microsoft from Google, aside from the hardware differences, is the way the company has positioned the technology. Google has offered their technology to a host of partners to develop across a number of discourse networks, from health to development to productivity. Microsoft has followed in the tracks of their previous office suite, operating system, and hardware, positioning their smart glasses for the labor force. In tandem with HoloLens 2, Microsoft is simultaneously advertising their Internet of Things software solution Azure. Azure is also being targeted toward the labor force as a way to connect "people, assets, processes, and systems" in an effort to "turn those insights into action with powerful applications built on the industry-leading platform for IoT development."[156] What this system establishes in relation to the HoloLens is an interconnectivity in which all actions and reactions are quantifiable. The quantification of information falls in line with current practices of productivity, data collection, and data processing. In other words, quantification helps propagate the culture of productivity that has become inherent in the contemporary workplace. This is especially true under the new Microsoft CEO Satya Nadella, whose vision and changes to the company have been in an effort to "help people be more productive."[157]

But productivity in the labor force is hardly a new concept. The optimization of a process or of a person through technology is also not born of the digital age. The optimization of people and processes belongs to a long history of labor and self-management. Why do we feel the constant need to optimize our ability to collect, store, and process information all in the name of productivity? Foucault defines it as follows:

> The practice of the self implies that one should form the image of oneself not simply as an imperfect, ignorant individual who requires correction, training, and instruction, but as one who suffers from certain ills and who needs to have them treated, either by oneself or by someone who has the necessary competence. Everyone must discover that he is in a state of need, that he needs to receive medication and assistance.[158]

What Foucault outlines is the need for improvement, an understanding that humans can always be better, and that if we are unable to improve ourselves through technologies of self-management, to look to others and the technologies of control that they offer. Foucault's explanation follows a long history of human development, and in this case industrial development. In 1880, a century before Foucault's publication *History of Sexuality, Volume 3: The Care of the Self,* Frederick Winslow Taylor began the field that would be known as scientific management. Taylor published *The Principles of Scientific Management* in 1911, following experimentation from his work within the trades from 1882 forward. The pursuit of the monograph and his overall experimentation examined ways in which to improve productivity within the trades and manufacturing. Taylor argues that "in the past man has been first; in the future the system must be first."[159] The production and use of spectacles falls within this pattern. Analog spectacles were a technology used to optimize individuals so that they could continue their labor past a specific age and in that way be more productive. Today smart spectacles are used to improve the system of labor where the individual becomes the means of action and, more important, replaceable. Smart spectacles function as a medium for the platforms that control information processing, and as such "they create a world where there is always the potential for something meaningful to be given, found, and delivered."[160] With the proliferation of smart technology, humans are shifted out of the circuit until they are left staring into it but are not a part of it.

When Jessi Hempel interviewed Alex Kipman about the Microsoft HoloLens in early 2015, she experienced a process that would become commonplace with the distribution of the HoloLens and Azure, as well as Glass and other smart glasses. With no electrical training prior to her task, the HoloLens was able to walk her through installing a light switch in a makeshift living room.[161] Pederson highlights the operability of smart spectacles in their ability to optimize productivity: "in one sense, the point of a head-worn wearable display is to unite thinking tasks (communicating) with moving tasks (embodiment) in ways that handheld devices are incapable of doing."[162] The device's ability to lead the user through a series of tasks with little to no error and an increase in productivity is a discourse many Glass Partners are using to explain why their productivity software on Glass is a need for industrial settings. Google itself claims that Glass is "rewiring productivity," an expression that follows the traditions of productivity seen in Taylorism, Fordism, and the LEAN model.[163] This is clear for the many partners of Glass, such as Picavi, GE, Upskill, Proceedix, and EyeSucceed, whose software is directed at productivity in the industrial sector.

Productivity is a central component of capitalism. It will continue to be made so through technologies of control, both digital and systematic. However, smart glasses are also playing a role in the reproduction of technology of control not often referred to in the case of capitalism: the panoptic gaze. Glass, HoloLens, and other smart glasses are a conduit for such a gaze. The reasoning is twofold. First, the digital or holographic information overlay comes from a central operating unit. Those who control the information operating unit control what the users see and therefore do, creating a central hierarchical system. Second, it simulates the panopticon gaze. When wearing smart glasses there is a level of information processing required to properly lay the digital information over it. The HoloLens uses sensors, Glass uses a camera, but in both cases what the viewer is seeing can be recorded and can be viewed elsewhere. When others can see everything that you see at any time, you behave as though you are always being watched. No longer only the subject of the gaze, humans have become the means for watching.

As subject and means of the gaze, humans are captured as their vision is captured as a means of perpetuating the relationship between power and knowledge. Which is to say, that by establishing tools to capture vision—whether through the analog spectacles that predetermine transmission of truth or through their smart and digital counterparts that not only control transmission but are also able to capture the gaze and modulate the subject—these dispositifs are able to keep power circulating through optical means. Within Kittler's conceptualization of the selection, storage, and processing of information, we see a fundamental shift of these information practices occurring outside of the human, and no longer within.

Glasnost

In 2013 this book's authors developed a prototype app for Google Glass called PerfectDay. PerfectDay worked with Glass's full-time visual and auditory data streams, compressed the data into manageable bits, stored all the data on our servers, and began to run machine learning algorithms to assess the arc of each user's day according to an ever-growing host of variables developed by our small but eager team. Eventually, patterns emerged that were both surprising and eventually impenetrable to our assessment. On the ground, however, by early 2014 PerfectDay was growing in popularity, especially among millennials as they attempted to transition from school and university to the workforce. PerfectDay proved particularly promising in terms of providing users the social management enhancements necessary for a disparate set of transient, mobile, and largely precarious set of employment milieus.

PerfectDay was simple to use and had one achievable goal: maximize the potential to make data-driven decisions for each and every moment throughout every single day. Past success would breed future success, but failure would also breed success. That is the beauty of PerfectDay. Unlike humans, who tend not to learn from their mistakes, PerfectDay has no such limitations. The ultimate goal was to make every day a PerfectDay. By leveraging the accumulated data of all PerfectDay users and modulating the importance of individuated user data according to how long any given PerfectDayer was synched, better and better choices could be made for nearly all situations faced. In simple terms, PerfectDay was an algorithmically driven decision engine that enhanced the capacity to become one's best self. Here's a simple account of how a PerfectDayer might benefit on a typical morning.

Just as most non-PerfectDay users sleep with their phones nearby, PerfectDayers always wear Glass, including at night. While monitoring slumber, PerfectDay assesses which songs or soundscapes produce the most appropriate emotional and intellectual state for the events on that day's calendar. At the most ideal moment, in the valley between REM cycles, but providing plenty of time to consume adequate and appropriate calories, update all social-media platforms, and go through optimal grooming routines before external demands call one forth, PerfectDay provides the stimulus to awake. On this day, a busy one filled with three different gig-shifts, a new PerfectDate first encounter, a telemeeting to plan a PerfectWeek retreat for some deep-data mindfulness mining, and a potentially painful meeting with parents to discuss student-loan payments, PerfectDay delivered something special (a song it saved for the most challenging days): Katy Perry's "Roar," the song that powered our PerfectDayer through her first heartbreak the year she became an early PerfectDay adopter. Arising at 7:22:34 provided the PerfectTime for heading to the kitchen where 172 milliliters of coffee awaited (any less and the 9:37 lag would kick in; any more and the social-media updates could veer into cat territory). The Google display provided notification that a two-egg, kale, banana, pomegranate, and carrot smoothie would provide superfood nourishment with just the right mix of protein and natural laxative; being regular is ever so important. Though the Timbits on the counter beckoned, when you looked at them PerfectDay provided a quick ImperVision of how the last morning eroded when you overrode its "no-Timbits suggestion." (You remember now. You were still overconfident, racing on sugar and wheat, and overrode PerfectDay's response to the question "What do you consider to be your weaknesses?" Totally not a PerfectDay!) "OK, fine," no Timbits this morning. Kale smoothie-me, please.

As you enter the shower and turn on the faucet, PerfectDay suggests 42 degrees Celsius, since anything hotter makes the skin on your calves dry out, and PerfectDay is quite sure you'll be wearing a knee-length skirt today. While you reach for your favorite shampoo (or so you think), PerfectDay mentions the humidity/wind/time-since-haircut ratio and flashes a quick ImperVision of the last time you used this product before a PerfectDate that didn't end perfectly. Clearly, the right shampoo can't hurt. So you grab the suggested shampoo among the group of five, squirt 11 milliliters into your hand, lather for 32 seconds, and rinse for 112 seconds. And then PerfectDay helps you with conditioner perfectly enhanced for humidity between 77 and 83 percent as long as winds stay below 10 kilometers per hour. Perfectly cleaned, no skin dried out or too oily, hair untangled but full of sheen you step out of the shower and grab the chamois towel, not the cotton. A scan of the closet provides a few more ImperVisions, but ultimately an outfit matched to this afternoon's PerfectDate's data-assessed desires, balanced against the need to get there on time (beware the high heels, which add four minutes to the walk and a 1.5 percent chance of a fall if one is hurrying), not spend money on an Uber (seeing as the likelihood of earning significant tips is low given the overload of staff this morning on a colder-than-usual day and your parents are assessing the joint debt account this afternoon as the next loan payment comes due), and then . . . "Who really has time to take all of these variables into account?," you say to yourself. PerfectDay answers: "I do."

Shortly, now fully optimized, you and PerfectDay head into public. PerfectDay begins its obligatory biometric optimizing scans of everyone and everything you encounter, assessing responses to each and every element of the PerfectYou they take in or are affected by. Where and for how long do others look? What expressive muscle formations run across their faces? Which muscles twinge and which fingers twitch in accompaniment with your own expressive data points? Noses are algorithmically micro-analyzed in a temporal slowdown and spatial magnification in order to assess odor responses to your olfactory transmissions. The micro-movements of skin and hair, especially around the ears, is used to assess what others hear.[164] Every data source, each engorged pixel collected by the miniaturized T-CUP camera, collects a "mind-boggling 10 trillion frames per second,"[165] which is then mined for potential meaning. This data-rich world is continually used to fine-tune your very own PerfectDay.

When two PerfectDayers pass on the street, a veritable data orgy begins as the circuit of assessment/counter assessment/response assessment kicks in and plays out at such a micro-temporal scale it's as if in stasis. A five-second encounter produces enough data to keep a circa 2020 server farm busy for years. Thankfully, Moore's Law failed to account for quantum computing. Like a reflexive infinity mirror, these moments are so data rich that PerfeConspiricists suggest that many such meetings don't happen by chance, even if it seems obvious that the outcomes will be less than perfect. As the day progresses and you settle into its perfection, each decision comes to you faster and easier. Desire aligns with App in something called PerfeSynchrony. As one of PerfectDay's most iconic ads suggested, "Now you don't have to worry about anything. Just let it go."

When Google Glass replaces smart phones, we will all be committed. Our personal psychiatric wards will be populated with the hallucinations befitting our needs. We will dwell there, not because we had been tricked or fooled, but because we will have subscribed. We will have subscribed because it will be best for us. Humans fail. They are faulty. They break down.

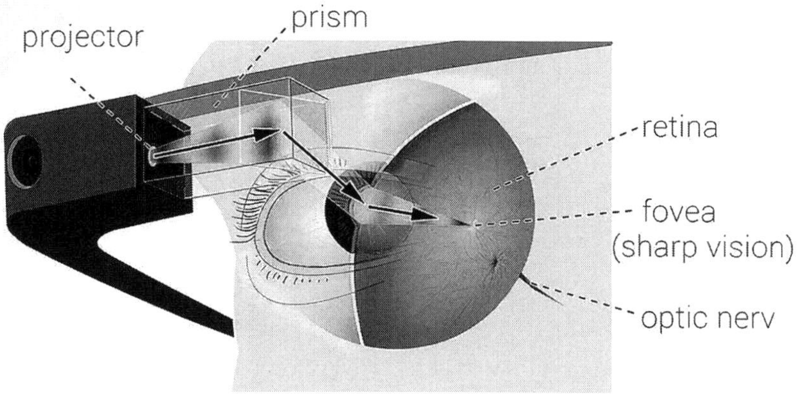

Figure 5.3. How Google Glass works. Image from Varifocals.net, Martin Missfeldt, spelling error in original.

They create substantial problems for others. This, we know, is the opening and ending to governance. We perceive this as truth, so why wouldn't we accept that the problems "we see" are real? When two circuits, the actually seen and the wished-for scene, are mutually perceived, one must overpower the other, lest they short-circuit, bringing darkness. In the end, hallucinations are the versions of events that are deemed incorrect. If a hallucination is created to be acceptable, to produce a desirable outcome, then it will be the truth. The truth game of the circuit is not truth to power, but rather power to truth. Such a purposeful psychosis will not be a problem, but rather the means to reconfigure the visual so as to solve all problematics. Once the visual is wholly malleable, the expansion of "prior knowledge" may robustly be used to reconfigure the very nature of perception, and hence governance and modes of subjectification will be radically reimagined. Once these hallucinations haunt all waking hours, eyes might as well be self-enucleated; for when what is seen has been digitally configured, all of one's life will be generated by the circuit.

6

Media Genealogical Method

The materials of communication cannot be considered separately from the power structures and subject positions that they (re)produce and co-constitute. We suggest that the best and most obvious means for doing this is to attend to the distinction that Michel Foucault introduced between *archaeology*, the term that best describes his work from the 1960s, and *genealogy*, which best describes the work he began in the early 1970s. The key difference of the later work was its investment in the analysis of power—or more specifically, technologies of power, technologies of governance, and technologies of the self. It is our goal to demonstrate that this difference *matters*, that it can inflect the entire body of media studies scholarship that follows Friedrich Kittler. Additionally, recognizing the archaeology/genealogy distinction helps unite media studies scholarship with other historical investigations into scientific, technological, and cultural practices that are already highly attentive to concerns of power/knowledge and subjectification, ranging from cultural studies to science and technology studies. Interest in anglophone countries of the overlapping arenas of German media theory, German media science, the work of Kittler, media archaeology, and cultural techniques is clearly growing. Yet, just as clearly, the terminology and canon used to explain and give credence to this work differ widely. We would like to simultaneously muddy and clarify these waters by injecting a new term into the mix: *media genealogy*.

This chapter offers theoretical and methodological demarcations of media genealogy. We begin by outlining the inherent limitations of the archaeological method that Foucault faced in the middle of his career and analyze his turn to genealogical method as a solution to these limitations. The second section traces the emergence of both British cultural studies and German media studies as alternate—and often antagonistic—responses to the Frankfurt School's hermeneutic critiques of mass media and culture. British cultural studies focused on polysemy and the slippage of meaning between encoded and decoded cultural meanings in mass mediated communications. German media studies, in contrast, focused on the technological constraints that determine the content of messages disseminated via media and communications technologies. We argue that British cultural studies preserves some aspects of Foucauldian genealogy but loses the benefits of strict archaeological analysis. German media studies, in contrast, advances archaeology by extending its scope of analysis to include nondiscursive technological artifacts while simultaneously neglecting genealogical focuses on power relations, subjectification, and change over time. In the third section we examine the way that German media studies has been positioned in anglophone discourses predominantly as "media archaeology." We argue that media archaeology intensifies the limitations already latent in German media studies by neglecting politics. We show how two recent turns in German research into cultural techniques and science and technology studies' research into the social construction of scientific facts and instrumentation can both be understood as useful extensions of media archaeology in a media studies discourse. In conclusion, we draw on the theoretical and methodological meditations of the chapter to articulate what media genealogy might look like.

Foucault: From Archaeology to Genealogy

As Colin Koopman has elaborated at length, even in *The Archaeology of Knowledge* Foucault was hounded by a dissatisfaction with his own theorizations of historical continuity in archaeological analysis.[1] At the time of writing, Foucault was already aware that he would be criticized for crudely affirming historical discontinuity. He explicitly acknowledges that "archaeology . . . seems to treat history only to freeze it," and later writes, "But, there is nothing one can do about it: several entities succeeding one another, a play of fixed images disappearing in turn, do not constitute either movement, time, or history."[2] He even closes the text by questioning the possibility of archaeologies other than those concerned with scientific epistemological practices, such as sexuality, painting, and politics. For him, analyses that remained too structural in nature "can never take place but in the synchronic cross section cut out from this continuity of history subject to man's sovereignty."[3] Foucault would soon come to believe that what all of his earlier works had failed to properly take into account was the problem of power, and they had so failed because the archaeological method had presented him with a synchronic snapshot of discursive rules, bracketing the need for an explanation of how those rules had emerged and endured across time. By 1970 he would posit the addition of a genealogical method to the archaeological method to help account for this; the two were originally meant "to alternate, support, and complete each other."[4] Here, as Hubert Dreyfus and Paul Rabinow (1983) are apt to point out, Foucault has begun to outline a methodology for articulating nondiscursive practices that effectively form a discourse.[5]

Foucault's lecture series at the Collège de France from 1970 to 1971 signifies a key turning point and provides great insight into this shift. Michael Behrent suggests that Foucault turns toward the Greek Sophists to accommodate his alternative to propositional knowledge (the outcome of archaeological method) with his interest in power.[6] Foucault's reading of the Sophists allows him to conceptualize power in terms of struggle rather than as something owned or that which is given through rights. Furthermore, power produces "truth effects—i.e. power-knowledge."[7] Finally, in Foucault's reading of the Sophists, discourse is material. Statements exist in time and space and occur through a medium. As such, statements cannot refer to objects; they are themselves objects.[8]

In addition to the Sophists, Foucault would repeatedly turn to Nietzsche as a means of elaborating genealogy. In his essay "Nietzsche, Freud, Marx" he elaborates the depth of archaeology that is viewed from higher and higher up "through" each respective philosopher's approach, producing a visible surface on which archaeological depth is laid out.[9] It is on this surface that later, in his essay "Nietzsche, Genealogy, History," Foucault locates the "nonplace" at which adversarial wills engage in the endless play of repeated dominations that leads to the emergence of forces.[10] These emergent forces are, in turn, responsible for the production, schematization, maintenance, inflection, and reproduction of the rules that constitute discursive regimes. As Foucault writes: "Rules are empty in themselves, violent and unfinalized; they are made to serve this or that, and can be bent to any purpose."[11] Their formation through the play of forces is anonymous, prior to the distinction of subjects and objects. Foucault writes, "Consequently, no one is responsible for an emergence; no one can glory in it, since it always occurs in the interstice."[12] At the level of the visible surface along which forces emerge and dominate one another, there are no identical points enduring across time, but the genealogist can isolate "substitutions, displacements, disguised conquests, and systematic reversals."[13] It is here that the genealogist can recover history as series of interpretations and practices[14] cutting across the temporal multiplicity of a field of forces.[15]

By the time *Discipline and Punish* was published, Foucault's new theories had matured to a much more stable state.[16] There he articulated a fully formed outline of power/knowledge, wherein knowledge is no longer isolated and is instead always coupled to practices and power dynamics. The mutual inflection of power and knowledge unfreezes time for Foucault, their intermingling having allowed for emergence in difference and repetition across temporal interstices. As Foucault notes, the ability to transverse the temporal axis in critique is not meant to better understand the past in terms of the present, but rather to produce *the history of the present.*[17] As Dreyfus and Rabinow explain, this new form of critique is able to locate the points at which "meticulous rituals of power" and "political technologies of the body" arose, took shape, and gained importance.[18] In so doing, Foucault finds a new way to offer up political opportunities to those who might seek them; his histories of the present delineate possible paths of attack, effects of truth ready for battle, in a struggle "waged by those who wish to wage it, in forms yet to be found and in organizations yet to be defined."[19]

For Koopman, problematization is the thread that unifies archaeology and genealogy under the banner of providing critique in the form of a history of the present.[20] For Foucault, problems emerge when a field of action, behavior, or practice becomes uncertain and unfamiliar or is set upon by difficulties imposed by (often nondiscursive) elements surrounding it (e.g., social, economic, or political processes). Around this problem, a number of possible solutions or responses are posed simultaneously, and these possibilities are conditioned by, but are in no way isomorphic to, their surrounding elements. Foucault writes: "This development of a given into a question, this transformation of a group of obstacles and difficulties into problems to which the diverse solutions will attempt to produce a response, this is what constitutes the point of problematization and the specific work of thought."[21] This notion of problematization, in conjunction with the concepts of temporal emergence and of power/knowledge produced by the cooperation of archaeology and genealogy, is the core of Foucauldian critique. Here "critique now becomes an inquiry into the conditions set by problematizations as they manifest in the contingent emergence of complex intersections of practice."[22] This mode of critique can track the play of forces, the contestations of power that have produced the space of possibility for contemporary practice. As Koopman notes, "The point is not to discern how the intentions of those in the past effectively gave rise to the present, but rather to understand how various independently existing vectors of practice managed to contingently intersect in the past so as to give rise to the present."[23]

In media studies and media history, the problematization approach has the capacity to not only (archaeologically) articulate the specific affordances and constraints of a technical media apparatus's functions of capture, processing, storage, and transmission[24] but also to (genealogically) articulate the clashes of power that resulted as multiple technologies were (counter)posed as potential solutions within a problematic field, and thus trace the emergence of a stabilized (socio)technical apparatus. Media problematization understood from a purely formalistic perspective would address what forms of noise in the system have been discovered/created that necessitate elimination. The historical specificity and centrality of media as constituting the means by which phenomena and elements of the world are approached as a problem space would be central. Thus, media technologies are more than the materiality of their machinic embodiments; rather, they are a method of systematically and repeatedly addressing a problem. Media address the certainty and regularity of signal processing, of reducing uncertainty and noise—or, in broader terms, "disciplining" signaling or, more consequently, epistemological practices. Media eliminate problems through the mastery of signal processing. Foucault's example for such practices comes from a French military manual that outlined a program for precise system control.

> From the master of discipline to him who is subjected to it the relation is one of signalization: it is a question not of understanding the injunction but of perceiving the signal and reacting to it immediately, according to a more or less artificial, prearranged code. Place the bodies in a little world of signals to each of which is attached a single, obligatory response.[25]

As we'll see later, for Friedrich Kittler, a foundational figure in German media studies, it is quite consequential that such an example has military origins.

British Cultural Studies versus German Media Studies

We might productively understand both cultural studies and German media studies as attempts to leverage the insights of Foucault to move past the rigid hermeneutical and structural analyses of media and communication represented in Continental Europe by the Frankfurt School and of mass communication and audience studies in the United States and Great Britain. These latter approaches had two defining characteristics that would cause the rupture with both cultural studies and German media studies. First, they tended to focus more on the content of the messages being disseminated in media and communication, subjecting them to semiotic and structural analyses to determine their meaning and implications for audiences. Second, and deeply related, they tended to understand media and communication as a unilateral process where the intentions of institutions and "great men" authoring content were received transparently by their audiences. Both of these frameworks for understanding media rely on a markedly modern, Enlightenment-informed understanding of what knowledge is and how knowledge works (namely, that knowledge is a product made in one place by the aforementioned "great men" and then transmitted to another, lower place via the aforementioned processes). Both also rely on similarly simple conceptions of power and the political. John Durham Peters has described this early framework as a model of communication whose teleological end point is telepathy, the maintenance of perfect isomorphism between authorial intent and received meaning.[26] As John Nerone has shown, taken together, these two characteristics paint a picture of cabalistic media institutions tugging on the puppet strings of entire populations, and scholarship in this vein tends to examine the "great men" who exercise control over the consciousness of historical populations and "man" the levers of time's passage.[27]

This outline is not meant to demean these earlier approaches, which offered early and insightful windows into processes of the centralized organization of national culture. It is meant more so to demonstrate the point of rupture from which new interdisciplinary discourses emerged and articulated themselves in response to more complicated understandings of the entanglement of knowledge production, sociopolitical power, and subjectification.

The Emergence of Cultural Studies

Cultural studies emerged in a rapidly changing post–World War II United Kingdom. After the war, cultural production increasingly moved from communal and local sites to centralized institutions orchestrated by the state or corporations via mass media, such as national radio, newspapers, and television. People started to develop more complicated social identities than the old working-class designation allowed for. They were thus less uniformly subjected to discipline and in its place were subjected to state hegemony and the culture industry.[28] This renegotiation of social identity was the subject of Stuart Hall and Tony Jefferson's edited collection *Resistance through Rituals,* which looked at how the fragmentation of the working class led to the formation of multiple subcultures that hybridized and repurposed hegemonic cultural forms to express themselves and oppose mainstream culture.[29] The approach became common in British cultural studies, which increasingly focused on the formation of subcultures as sites of class-based politics.[30] An often-localized ethnographic approach was quickly expanded to analyses of mass communication media, such as David Morley's *The "Nationwide" Audience,* which deconstructed the "mass" audience for a nationally broadcast British news program to demonstrate how different segments or even individual members of that audience interpreted the program differently.[31]

As can be seen in the examples above, the British approach required a new theory of media and communication that afforded more agency to the "masses" that consumed messages as readers and/or audiences. The first movement toward this new theory was an acknowledgment of the polysemy of any signs contained in media or communications content. In essence, any given sign, such as a word or an image, always had multiple meanings, because most signs were not indexical (i.e., they did not point to specific objects in the world). Instead, they got their meaning from one another in terms of their positioning in a sequence and their selection in lieu of others—what semiotic analysis would call syntagmatic and paradigmatic relations to other signs, respectively. Polysemy opened up space for agency on the receiving end of messages, such that the mediation and communication of culture could then be examined in terms of hybridization, repurposing, remixing, negotiation, and play. As Simon During notes in his Introduction to *The Cultural Studies Reader,* "a concept like 'hybridization' still does not account for the way that the meanings of particular signifiers or texts in a particular situation are, in part, ordered by material interests and power relations."[32]

Already in its move toward polysemy, British cultural studies was indebted to French theorists like Roland Barthes and Michel de Certeau, who had written similarly, though in a less explicitly political way, about how culture was radically open to creative interpretation and only had meaning within the always-changing referential networks of discourse. To address the materiality and contestations of power that grounded this polysemy, though, British cultural studies would have to turn to other French thinkers of the era, most notably Foucault and Pierre Bourdieu. While Bourdieu's theories were most directly concerned with the typical cultural objects that the Birmingham School preferred to analyze, it is arguably Foucault who had the deeper and more persistent impact on cultural studies as a whole. While British cultural studies borrowed heavily from French thought, it certainly also hybridized, repurposed, and remixed what it borrowed to create highly original and practically applicable theories of culture, media, and communication.

Take, for example, Hall's paradigmatic piece "Encoding/Decoding."[33] In this piece, Hall argues that we ought to understand communication as a cyclical process—consisting of production, circulation, distribution, consumption, and reproduction (see Figure 6.1)—rather than as a unilateral dissemination. Hall spends the first half of the article refuting the standard behaviorist tenets of audience studies in mass communication by pointing out how the polysemy involved in communication and the performative labor of interpretation necessary for decoding messages ensure that there is never perfect isomorphism between encoded and decoded meanings. However, once Hall is comfortable that he has sufficiently debunked this approach to studying mass communications, he returns to a point he made earlier in the article: that this communicative process can best be thought of as "a 'complex structure in dominance,' sustained through the articulation of connected practices, each of which, however, retains its distinctiveness and has its own specific modality, its own forms and conditions of existence."[34] In short, there are preferred meanings that constrain both the encoding and decoding of any transmitted message. These preferred meanings correspond to institutional, political, and ideological power

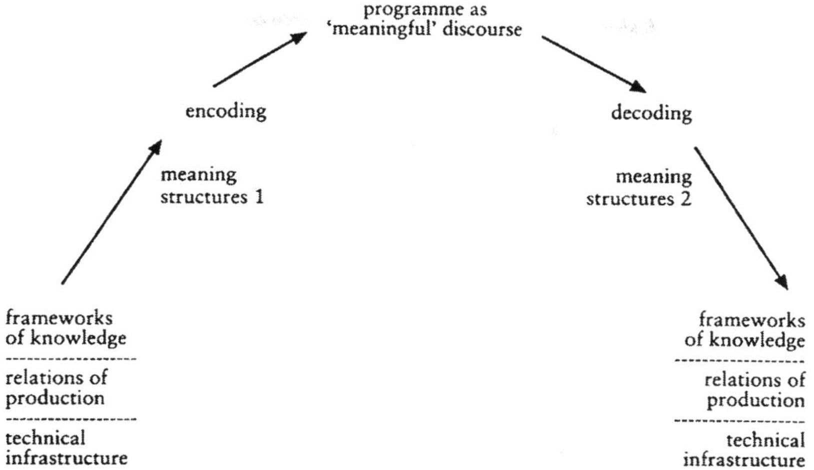

Figure 6.1. Stuart Hall's process of communication. Stuart Hall, "Encoding/Decoding," in Culture, Media, Language: Working Papers in Cultural Studies, 1972–79, *ed. Stuart Hall, Dorothy Hobson, Andrew Lowe, and Paul Willis (New York: Routledge, 1980), 120.*

structures. These power structures are in turn engaged in a constant and distributed struggle to "enforce, win plausibility for and command as legitimate a *decoding.*"[35]

What we can see here is Hall astutely pushing Foucauldian analysis into the terrain of mass media. Rather than looking at mass media as unilateral dissemination mechanisms for homogeneous and homogenizing messages (à la the Frankfurt School), Hall understands the encoding and decoding of mass mediated messages to be the grounds of contemporary power struggles over not only what interpretations are possible but, given the cyclical nature of his communicative process, what future messages can be produced and encoded in the first place. This added layer of complexity is essential for understanding how mass media play a role in subjectification, which is never a unilateral process, but instead one of negotiation and hybridization brought on by polysemy and the non-isomorphism between encoded and decoded meanings. In short, Hall politicizes the analysis, and does so in a way that produces space for resistance. The weakness of Hall's model is that he understands the formal dimension of mass mediated messages to be entirely *discursive,* and looks only to the contestations over semiotics to understand how the encoding and decoding of messages gets inflected and constrained. He pays almost no attention to the technical infrastructure at either end of his model, which certainly has as large a role in constraining the encoding and decoding of messages as discourse. It is this critical gap that German media studies will work to fill, but as we'll see, it often loses the political potency of Foucauldian analysis in its archaeological digs for technical specificity.

The end result is that cultural studies emerges as a discourse uniquely positioned to examine subjectivity—both as a hybridized and negotiated production through the agential interpretation of cultural messages *and* as a radically individualized perspective on the world rooted in everyday life. As such, cultural studies is by default a *politicized* form of analysis, as it is constantly revealing the ways in which power intersects with and cuts across the individual bodies and identities of subjects.[36] This focus on power has allowed cultural studies to keep pace with the times and spread across the globe. It was center stage in the criticisms of Thatcherism (and Reaganism), and since has productively hybridized its own takes on feminism, race theory, postcolonialism, sexuality and queer theory, cosmopolitanism, globalization, transnationalism, science and technology studies, and affect theory. Its commitments to subjectivity and politically engaged scholarship have given it global purchase; there is no question that cultural studies ought to be engaged by future media studies.[37]

Cultural studies' framework for power is similarly well suited for the twenty-first century. For cultural studies, there is no central agency directing all of society unilaterally such as there was in traditional twentieth-century critiques of capitalism or ideology. Resistance becomes tactical, fragmented, and occurs in the cracks and gaps, rather than via large-scale strategies orchestrated at the national or international level.[38] This is another key strength of cultural studies for work in media studies and media theory. We can see the new lens of distributed power if we return to Hall's model of communication, where his theory of structures in dominance and performative interpretation offers us a strong analytical tool for examining how encoding and decoding are inflected by the field of possibilities that results from current, local contestations of power. On the other hand, the critical weakness of cultural studies is that it has little to say about the middle prong of its model of communication.[39] There is no rigorous theory of *mediation*. That is to say, Hall ignores the way that technological affordances and constraints can inflect encoding and decoding processes. As we'll see below, this is the key strength of German media studies. Cultural studies, then, is fully engaged with the genealogical arm of Foucault's research and has continually expanded and reimagined it to suit new conjunctures of subjectivity and power, but it hasn't made many advances with his archaeological method. It remains bogged down in semiotics and discourse analysis. To phrase it reductively, cultural studies is good at genealogy and bad at archaeology.

The Emergence of German Media Studies

Geoffrey Winthrop-Young has argued that the ethnic diversity and deeply rooted class systems in Great Britain led to cultural studies' focus on subversive interpretation and differentiation—particularly what he terms "mechanisms of vertical differentiation."[40] In contrast, Germany has had a less entrenched class system, much more rapid industrialization, fewer and more forgotten experiences with colonialism, and has been organized through the production of a cultural canon and linguistic affinities that instill a collective identity. Per Winthrop-Young, "Germany, to put it bluntly, is a kind of media product."[41] As such, Germany has historically been much more inclined to focus on homogenization rather than differentiation, looking at the production of (national) identity as a cultural leveling carried out by media and communication.[42] This is evidenced both in the Frankfurt School's model of the culture industry *and* in German media studies' focus on media as the a priori determinants of culture and politics. The difference is that German media studies attains this focus on homogenization and cultural leveling by examining the technical specificity of media and communications technologies, arguing that technical dimensions of media apparatuses constrain discourse a priori. In so doing, they push Foucauldian analysis into new frontiers by analyzing the *nondiscursive* dimensions of power and governmentality.

Take, for example, the patriarch of German media studies, Friedrich Kittler. For Kittler, media technologies certainly arose within the confines of discursive power relations, particularly in the nineteenth and early twentieth centuries. These discursive conventions constrained what technologies could be imagined, theorized, sketched, planned, and produced in myriad ways. For example, media were originally only imaginable within the confines of individual sensory channels and as a result of psychological and physiological research that was able to increasingly mathematize human sensation.[43] Some media, like the phonograph, required no raw materials that were not available in antiquity, and thus could have been invented at any time had the discursive conventions rendered them imaginable.[44] Others, like photography, required not only to be rendered imaginable, but required entire other discourses to arise, like chemistry for the production of silver nitrate.[45] While Kittler often uses the shorthand of inventors' names, he is careful to note that technical media are always the result of a chain of discursive and technological shifts, including a number of important failures and dead ends, that are pursued across time and space by various teams of people who not only invent, but develop, organize, improve, and alter them.[46] In short, "Because a human life is far too short to comprehend avalanches of technical innovations, teamwork and feedback loops become essential."[47]

At this point, Kittler's Foucauldian influence is clearly showing, as he maps the discourse networks distributed across time and space that led to the emergence of some technologies rather than others—in slight contrast to Foucault's question of how one particular statement was utterable rather than another. Kittler often acknowledges that he borrows heavily from Foucault for his method of historical analysis. It follows, then, that a "discourse network," Kittler's media-oriented, Foucault-derived term for an epoch, comprises "the network of technologies and institutions that allow a given culture to select, store, and process relevant data" and that an "archaeology of the present must also take into account data storage, transmission, and calculation in technological media."[48] The "also" here refers to the fact that Foucault's discursive formations "did not progress beyond 1850" and hence paid no attention to the second Industrial Revolution and "its automation of the streams of information."[49] Thus Kittler seeks to historicize not the processes of meaning making, ideological struggle, or cultural reproduction, rather his work "follows the Foucauldian lead in that it seeks to delineate the apparatuses of power, storage, transmission, training, reproduction, and so forth that make up the conditions of factual discursive occurrences. The object of study is not what is said or written but the fact—the brute and often brutal fact—that it is said, that this and not rather something else is inscribed."[50] This posthermeneutic approach to media history appeared as a corrective to the prevailing assumptions and concerns of literary studies, Kittler's initial academic home. The key here for Kittler is not meaning production, the play of ideology, or idealized discourse communities and publics. His maneuver out of the hermeneutic tradition depended upon a decentering of human agency and meaning production via a turn toward Foucauldian discourse analysis and Claude Shannon's model of communication.

In large part, Kittler's analyses are indeed archaeological in nature. He is often focused on demonstrating exactly how particular technologies function in relation to the "so-called human"—or, as Kittler terms it, how they escape the grid of the symbolic. These analyses are of specific inventions leading up to and including phonographs that digitally capture the real, or photo- and cinematographic devices that digitally capture the imaginary. And while any given device's technological functioning—in terms of things such as component arrays, wiring, hardware, and programming—cannot be explained by their original frames of reference, *their emergence can.* While Kittler does trace specific technologies out of joint with any linear technological history, he is also invested in analyzing their disjointed emergence across the interstices of multiple temporalities. He does so by producing numerous overlapping genealogies, as in his work on optical media, which traces a lineage from the camera obscura to photography to cinema to television, which affords him theories of optical mediation as a set of nondiscursive practices and techniques. These crystallize into variable media technologies and, we suggest, operate as modes of subjectivation under what Foucault came to call governmentality in the late 1970s. It is here that we note a particularly productive line of inquiry into the relationship between media archaeology and governmentality that has largely been overlooked.

One of Kittler's primary framing mechanisms is the argument that differentiating technical media based on the human sensory channels they correspond to is arbitrary, for there are only multimedia systems. Our capacity to differentiate them by such a correspondence only exists because "they were developed to strategically override the senses."[51] Once media channels become digital, they all become interoperable and interchangeable. At that point, technical media render the human central nervous system superfluous to mediation.[52] The human is lost into the apparatus, and the so-called human is split into physiology and information technology. At this point it is clear that Kittler is moving past the archaeological method. His work is concerned with tracing the emergence of particular technical media, and it does so in relation to the play of forces forming discursive rules, governmentality, strategies of producing bodies, and technologies of the self that operate under the regimes of physiology, anatomy, and psychometrics—particularly in relation to military endeavors.

Technical media arise out of distributed discursive networks composed of bodies of knowledge (such as chemistry and physiology), objects of knowledge (such as precise chemical components), distributed human research and development teams, techniques and technologies of governmentality and the self, and recombinations of technological apparatuses. After crystallizing, technical media quickly cut their roots and become mobile, intermixing and recombining with one another, to the point where their original frame of reference holds explanatory power for their functioning.

In his earliest work, Kittler describes his project as an analysis of "the network of technologies and institutions that allow a given culture to select, store and process relevant data."[53] We would like to highlight these aspects, at times neglected, of Kittler's work. Although many view the termination of Kittler's history to be the obsolescence of the human in a world of digital computers and near autopoietic machines, we might take as a clue the sparse amount of his work that actually elaborates that world. Instead, one might see Kittler as diagramming a particular problematization in the contemporary, an arbitrary and contingent set of possibilities for the future in our *current* discourse. Despite the emphasis he places on it, the world Kittler imagines never arrives in his work in any teleological sense. If we take that to be true, it is easy to envision Kittler as writing a history of the present, with an eye to one possible future that has not yet arrived, so that we might alter any of the multiple strategies and tactics plotting our course. Jussi Parikka seems to agree that, in Kittler, at the birthplace of media archaeology, his two key insights—media as "systems for transmitting, linking and institutionalizing information" and the functioning of power in an age of technical media—are the result of a combination of both archaeological and genealogical methods.[54] Perhaps this is why Kittler eschewed his supposed affiliation with media archaeology.[55]

In some ways it is rather obvious. Kittler's view of human technological interaction feels cold and unflinchingly harsh.[56] Much of the material he worked with was firmly situated in the roots of a German intellectual and literary tradition that is not easily penetrated by those unfamiliar with it.[57] Further, Kittler draws not only from Foucault but also quite heavily from Marshall McLuhan and the French psychoanalyst Jacques Lacan. Ensuring it is even less palatable upon first review, Kittler's fundamental theory of communication is borrowed from Shannon and Weaver's classic "transmission model," which has in the United States been situated as an adversary to critical and cultural approaches following James Carey's

1975 article, "A Cultural Approach to Communication."[58] Finally, Kittler's historical teleology may seem problematic. All media technologies are viewed in terms of their perceived similarities with digital media and/or what technological, mathematical, economic, or theoretical elements they added to help "media history [culminate] in the digital computer."[59] Stone tablets, ink quills, and typewriters are necessary, though ultimately failed, attempts to reduce the noise that kept humans from hearing the sirens' call for a digitized world.

A further ingredient for Kittler isn't so much methodological, but is rather his dependence upon Canadian Medium theory. Combining Harold Innis's concern with space time bias with Marshall McLuhan's "extensions of man," Kittler looks historically upon the specific means by which time shifting and spatial transmission works to extend the sensory, memory, and logical capacities of the human.[60] Whether this be by book, telegraph, television, typewriter, or computer, the brute fact of mediation according to Kittler is best characterized by Shannon's classic mathematical model (much misused and maligned in the United States by perspectives that take it out of the context of its inception) by which messages (or statements) are encoded by a sender, are transmitted through a medium, reach a receiver, and are decoded. Shannon's theory in the hands of Kittler elaborates the processes by which noise is overcome in order for data to be stored, transmitted, and processed. Thus, Kittler's Foucauldian media history, most succinctly presented in *Optical Media*, is the history of technological achievements in which the camera obscura and lanterna magica are not merely devices that predated photography and film, but rather elements in discourse networks that reshaped how optical data were collected, moved about, stored, and processed. Such media technologies reorient power/knowledge formations as they create new means for knowing and acting upon the world. For instance, the camera obscura functioned mimetically as a means for an understanding of human optics, perspective, and even ballistic predictability. Further, from such a vantage, media are certainly "extensions of man," but only insofar as the "man" in question was never an a priori, but rather an always-already-extended product of discourse and technology. Thus, McLuhan's humanistic sensibility gets reworked.

There are two specific takeaways from Kittler's work for thinking about Foucault and media history. First, the study of discourse as it relates to media looks very different from what Kittler derides as a "trivial, content-based approach to media."[61] Rather, discourse relates to the networks of associated technologies and statements that organize what can be said (data-knowledge), where and to whom it is said (transmission-power), and what can be made with such statements (processing-power/knowledge). The importance of media technologies is that they changed such dynamics. They made the visual and auditory realms mechanically capturable, storable, searchable, transmittable, and processable. They combined the time shifting of memory or print with analog data capturing that did not have to pass directly through human observation, but rather through mechanical means of production and processing. Cameras may have first depended upon human hands to be made operable, but once freed from such constraints they are capable of capturing data regardless of intentionality, human agency, or the constraints of meaning making. Much meaning will be made and humans will enact agency via the camera for all sorts of intent, but the pivotal issue for Kittler is how it transforms the capturing of data. This is one aspect of the facticity of Kittler's media analysis. The world was being newly inscribed, not by human hands writing, chiseling, or making marks on papyrus, paper, or some other surface (what Kittler refers to as the artisanal stage of media production), but rather through apparatuses that combined chemistry, mechanics, and optics.

Sybille Krämer suggests that Kittler's media history is fundamentally about *both* technologies *and* institutions:

> It is necessary to note that there are not always data, on the one hand, and then, on the other hand, the media that are concerned with the data. It is far more the case that media are the production sites of data. These production sites are discourse systems, the networks of techniques and institutions that preprocess what will even be considered data in a given epoch.[62]

Here we have one possible means for overcoming the divide that Nerone sees in the field of media history between examining technologies and institutions.[63] Because Kittler's historical accounts are so attentive to the inner workings of technological development, we are given a view into the institutional laboratories from which they are formed. From such a vantage, specific techno-mechanical, biological, chemical, electrical, and human assemblages are the actors, institutional demands and physical limitations are the well-lit stage.

Second, and most central to a Foucauldian media history, may be the question of knowledge production and the games of truth that are increasingly played out via media. Media are fundamental to knowledge production; from how data are collected, how they are made visible, their form, the life of their existence, their degree of malleability, the extent to which they can be translated from machine to machine to machine, and ultimately how they can be processed to make things happen. In the end, for Kittler the digitalization of data, of the world, of knowledge makes all media "the same." From such a vantage, the effects on the continuing production of knowledge have clearly not fully come to be understood. In terms of a Foucauldian media history, the project of making sense of how media have altered the production of knowledge that organizes how the world is known, processed, and acted upon has barely begun. Yet what is needed isn't merely the Foucauldian project of archaeology (of discourse), but rather of genealogy (of media and its emergent discourse).

Kittler's Foucault needs to be complicated by power relations to counter what Peters calls his "disdain for people, or more specifically, for the category of experience."[64] In particular, we would like to imagine that many of Kittler's insights would not lead to categorizing humans into what emerges in his work as (1) the geniuses of scientific invention and philosophic understanding or (2) the great masses of unnecessary and powerless "so-called humans" who fail to populate his histories.[65] Rather, a Foucauldian media history might focus not only upon the early Foucault of knowledge production and discourse networks with their attendant ruptures, but more so on Foucault's later work on power and subjectification. Our understanding of the relationship between subject formation and media would become richer and more complex, not flatter and reductive. This would involve focusing not only upon Kittler's favored loci of struggle, nationalist scientific competition, and war, but rather on the messiness of struggle over epistemology and invention that takes place in local contexts of lived experience. The give-and-take between from-on-high and down-low has been a difficult relationship to chart. The archive favors the memory, and hence the perceived power, of those who could write their thoughts, file their notes, and mail in their patent formulas. At the very least we know that the creative power of "the people" to create their own discourse networks has been alive and well for a very long time. Foucault summed up the two approaches in his 1975–76 lecture series *Society Must*

Be Defended: "To put it in a nutshell: Archaeology is the method specific to the analysis of local discursivities, and genealogy is the tactic which, once it has described these local discursivities, brings into play the desubjugated knowledges that have been released from them."[66] So where might we find such release? We can start up high, before moving below.

The Current State of Media Studies in Anglophone Discourse

German media studies is very concerned to get the science and technology right, to have some grounding in the actual circuitry of the electronics being critiqued. While this is certainly not the entire thrust of German media studies, it certainly is the predominant feature in its translation for English-language audiences who mostly utilize German media studies under the banner of "media archaeology." In Erkki Huhtamo and Jussi Parikka's edited collection *Media Archaeology: Approaches, Applications, and Implications,* which for many served as the first English-language aggregation of media studies scholarship operating under the banner of media archaeology, the editors describe Foucault as "formative" for many media archaeologists and as the most prominent forerunner of "media-archaeological modes of cultural analysis."[67] In fact, most scholars doing work that might be described as media archaeology are influenced by, directly cite, and work through the central concepts of Foucault.[68] Across the literature, it seems as if the archaeological component of media archaeology is always Foucauldian in its origins.

Parikka has done the most extensive work to aggregate media archaeological scholarship as well as to elaborate the theories and methodologies most typical to a media archaeological approach. He argues that Foucault's influence is largely methodological and utilized for "excavating *conditions of existence*" for a given object.[69] He writes: "Archaeology here means digging into the background reasons why a certain object, statement, discourse or, for instance in our case, media apparatus or use habit is able to be born and be picked up and sustain itself in a cultural situation."[70] In his most extensive treatment of Foucauldian methodology, Parikka cites *The Archaeology of Knowledge* exclusively, even paraphrasing what he understands to be the key methodological principles of the text.

This leads us to point out an opening for a further investigation into the potential value of media genealogy: first, archaeology as a practice specific to media could be more thoroughly accounted for in order to elaborate its limitations, and second, a cursory review of the projects operating under the media archaeology banner demonstrates that they often overstep the methodological limitations of archaeology and begin to operate in a genealogical mode that is unfortunately under-recognized as such. We see these tendencies in Parikka's original work on the intersections of media and nature.[71] In his book on geology and media materialism, Parikka attempts to construct "a creative intervention to the cultural history of the contemporary"[72] by looking beyond the internal specifics of the machine and instead taking machines as "vectors across the geopolitics of labor, resources, planetary excavations, energy production, natural processes from photosynthesis to mineralization, chemicals, and the aftereffects of electronic waste."[73] As demonstrated later in our examination of the limitations to the archaeological method, this is a research agenda that seems genealogical through and through. We find the same tendency in Parikka's work on insects and bestial media archaeology, where he hopes "to look at the immanent conditions of possibility of the current insect theme in media design and theory,"[74] which would seem to imply critique by problematization, something we suggest is central to media genealogy.

Parikka explains: "Media archaeological methods have carved out complex, overlapping, multiscalar temporalities of the human world in terms of media cultural histories."[75] It is our contention that this is precisely what media archaeology *cannot* do without turning toward genealogical practice. Our critical attention on Parikka is not meant to detract from the original and important body of work he has brought to the English-speaking world, but instead to more prominently open up the field of Foucauldian-influenced media studies to genealogical method, specifically through the notion of problematization. As the herald of post-Foucauldian German media studies in the English-speaking world, Parikka's work offers a convenient entry point into such an expansion, and we call for an alignment of goals for future media studies scholarship. Toward that end, we will examine two current discourses that have significant overlap with what we envision as media genealogy, those of culture techniques and of science and technology studies. After reviewing current research into cultural techniques and the social construction of scientific knowledge and technological artifacts, we establish some of the possibilities that media genealogy and problematization offer to media studies scholarship and sketch out a media genealogical methodology.

Cultural Techniques

As Peters has argued, Kittlerian media theory relies on a universalization of "Western" culture and is unable to address the most pressing political and ethical stakes of media and technology.[76] These limitations are at least partly addressed in the more recent German championing of *Kulturtechniken* as a future avenue for media studies. Bernard Geoghegan said of this new direction in media studies that "this is not *media archaeology* but rather an archaeology *of media*," and he has argued that "this overturns the anti-biologism that prevailed in nearly all Kittlerian analysis and points towards a genealogical complement or alternative to media archaeology."[77] In essence, cultural techniques are increasingly positioned as the solution to the limitations of media archaeology. As we'll see, they broaden the focus of media analysis beyond technological artifacts to include communities of practice, relations of power, and technologies of the subjectivate individuals.

Kulturtechnik can be translated as "rural" or "environmental engineering." It has roots in the Latin *cultura* and *colere,* which give us the English term *cultivate.* These terms thus originally relate primarily to agriculture. However, as Winthrop-Young notes, "agriculture . . . is initially not a matter of sowing and reaping, planting and harvesting, but of mapping and zoning, of determining a piece of arable land to be cordoned off by a boundary that will give rise to the distinction between the cultivated land and its natural other."[78] This is precisely the sense in which the term was first used by Friedrich Wilhelm Dünkelberg, who in 1876 introduced a course on *Kulturtechnik* for the study of land improvements to promote agriculture.[79] For nearly a hundred years this was the primary sense of the term. It was only in the 1970s that it was taken up to describe "the skills and aptitudes involved in the use of modern media."[80] The new concern was with the conceptual and technological skills required to engage with mass media, like television, which required viewers to be able to interpret all sorts of audiovisual content.

This new media-centered conceptualization of *Kulturtechniken* was quickly expanded. As Geoghegan notes:

> Cultural theorists at the Humboldt University of Berlin (e.g. Christian Kassung, Sybille Krämer and Thomas Macho) identify cultural techniques with rigorous and formalized symbolic systems, such as

> reading, writing, mathematics, music, and imagery. . . . Researchers in Weimar, Siegen, and Lüneberg tend towards a more catholic definition that recognizes a broader range of formalizable cultural practices, including tacit knowledge, and class-laden rituals of Victorian servants, and the law as cultural techniques. . . . Binding together these varied definitions and understandings of *Kulturtechniken* is a shared interest in describing and analysing how signs, instruments, and human practices consolidate into durable symbolic systems capable of articulating distinctions within and between cultures.[81]

As Winthrop-Young has noted, this expansion of the domain of *Kulturtechniken* has increasingly linked the concept to embodiment and subjectivity. Writing, for example, requires a long process of disciplining the body in terms of posture, grip, penmanship, eye movements, and attention. By immobilizing our bodies and focusing our minds, we gain the conceptual mobility of navigating diverse texts. This newer understanding of *Kulturtechniken* draws heavily on Marcel Mauss's earlier theories of *techniques du corps,* where he analyzed walking, swimming, marching, and trench digging as bodily techniques each with particular styles ingrained through disciplinary apparatuses.[82]

Research into *Kulturtechniken* thus starts from a rudimentary assemblage of techniques, practices, instruments, and institutions that precede both their stabilization into increasingly autonomous technologies *and* the concepts generated to interpret, analyze, and classify them.[83] *Kulturtechniken* are thus essential "technologies of the self" that help produce and maintain our identities.[84] Further, as Bernhard Siegert notes, "The concept of cultural techniques highlights the operations or sequences of operations that historically and logically precede the media concepts generated by them."[85] Or as Winthrop-Young phrases it, "Cultural techniques are preconceptual operations that generate the very concepts that subsequently are used to conceptualize these operations. Emergent phenomena turn into their own foundational properties."[86] This constitutes just the sort of temporal paradox of the Foucauldian disjuncture. As Winthrop-Young puts it, here the chicken lays the very egg from which it hatched.[87] The most frequent example of such a cultural technique comes from Siegert's work on doors. For Siegert, the closest thing to an *ur*-cultural technique is that of species differentiation.[88] The differentiation between humans and other animals comes about through the mediation of the cultural technique of doors, which create domiciles, fences, pens, and corrals, the necessary operative thresholds for enclosing space and creating both interiority and exteriority. It is from this operation of enclosure that humans and other animals differentiate themselves in relation to one another and become domesticated. This differentiation then comes to ground all relations between humans and other animals.[89]

It is essential to read this through the chicken/egg paradox to grasp the uniqueness of the approach. Here humans and other animals literally become themselves through the mediation of doors as cultural techniques only to subsequently envision themselves as the agents that construct and operate those doors. As Winthrop-Young notes, *cultural technique* "refers to operations that coalesce into entities that are subsequently viewed as the agents or sources running these operations."[90] It is worth noting that this approach was already in embryo in Siegert's earlier work on postal delivery apparatuses that by situating subjects in a predetermined postal grid erase the contingencies of sender and receiver positions to instead constitute singular postal subjects in a closed circuit of postal circulation.[91] For Siegert, the focus on cultural techniques moves away from both hermeneutic and ontological analysis toward a historical investigation of the "chains of operations that link humans, things, media and even animals."[92] Just as we've seen in Foucault, Siegert argues that the only fundamental continuity of the world at an ontological level is an abyss of non-meaning that is constantly hidden by the historically conditioned emergence of differentiations of space, time, subjects, and objects.[93] It is worth noting that this also implies that there is no singular Culture, but only plural cultures.[94]

The movement toward cultural techniques deprivileges technological artifacts, moves away from hermeneutics and ontology, and instead focuses on the historically conditioned chains of operations that simultaneously articulate subjects, objects, space, time, and meaning. This, then, is a version of media studies that retains the capacity to critique both the nondiscursive technical/technological mediations of power/knowledge *and* broader cultural practices like subjectification. We might read this as a reinvigoration of the Foucauldian dimensions of media studies. This move is evidenced clearly in the writings of scholars of cultural techniques. As Siegert notes, "research into cultural techniques cannot be conducted without an analytics of power."[95] As Geoghegan notes, the uptake of Kittlerian analysis led to a conflation of all media and a loss of focus on culture, technique, science, and broader power relations.[96] For Geoghegan, the renewed focus on cultural techniques gets media studies past this impasse by pushing our methods from archaeology toward genealogy. He writes: "No media archaeology offers a resolution to this dilemma. Instead, media genealogists must ask how, and under what conditions, cultural techniques strategically and temporarily consolidate these forces into coherent technologies."

Science and Technology Studies

The German move to cultural techniques shares a lot of similarities with actor-network theory (ANT), particularly the work of Bruno Latour, as well as broader science and technology studies (STS). As Winthrop-Young notes:

> The similarities to the ANT approach (as well as to the work of Bruno Latour) are obvious: procedural chains and connecting techniques give rise to notions and objects that are then endowed with essentialized identities. Underneath our ontological distinctions are constitutive, media-dependent ontic operations that need to be teased out by means of a deconstructive maneuver able to disentangle acts, series, techniques, and technologies.[97]

ANT—and STS more broadly—refuses to treat the history of science and technology as a history of ideas, of essential forms that manifest themselves in the world in correspondence with the mythical narrative by which science uncovers the real world. Instead, these traditions look at the discursive and nondiscursive material constraints of the problematization that inflects what can and cannot be discovered, what can and cannot be invented. As such, we might read them as being not so much historical as *genealogical.*

The emergence of STS paralleled the theoretical synthesis that produced cultural techniques in Germany, and it derived from similar exigences. To oversimplify: the groundbreaking perspective of American physicist-turned-historian Thomas Kuhn (*The Structure of Scientific Revolutions,* 1962), informed by the work of Polish biologist-turned-philosopher Ludwik Fleck (*Genesis and Development of a Scientific Fact,* published 1935, translated 1979), established that the knowledge-producing practices of the hard sciences were also social practices, which opened the history of science to the post-positivist lenses of sociology and anthropology. Interestingly, Foucault published *The Birth of the Clinic: An Archaeology of Medical Perception* in 1963, just one year after the publication of Kuhn's *The Structure of Scientific Revolutions,* so Foucault was arguably "doing" STS before the field existed. No rigorous history of science after 1962 could ignore subjectivity.

One major ingredient in the conceptual glue that helps to synthesize the social and the material in STS accounts is Foucault. Like media studies, genealogical STS thinks with Foucault to move past rigid conceptual and methodological frameworks of history of science in order to see from outside the scientific knowledge-making apparatus. And a right-minded genealogy already sees the material (media, in many cases) as constitutive of social relations. Much of Foucault's early anglophone uptake, however, leaned on the subject/power dynamic and neglected the archaeological piece. Latour says of Foucault: "No one was more precise in his analytical decomposition of the tiny ingredients from which power is made and no one was more critical of social explanations. And yet, as soon as Foucault was translated, he was immediately turned into the one who had 'revealed' power relations behind every innocuous activity: madness, natural history, sex, administration, etc. This proves again with what energy the notion of social explanation should be fought."[98]

Since the 1970s and 1980s, then, the social study of science (SSS) in Europe and the Kuhn-informed history of science, rhetoric of science,[99] and sociology of science in the United States (all of which we will refer to broadly as STS moving forward) were analogous to the cultural studies of Hall et al. that we have described earlier in this chapter. In their understanding of the cultural openness of scientific practice, STS often shares with cultural studies a focus on attending to both the epistemological and the political stakes of scientific discovery and technological invention.[100] In the best of cases, they can weave together what Donna Haraway has described as the "situated knowledges" of the marginalized to produce a bottom-up consensus in lieu of the phallic omniscience of scientific objectivity—itself a social construct and thoroughly dependent upon media technologies, as Lorraine Daston and Peter Galison have shown in their arguments about the birth of "mechanical objectivity."[101] It is precisely this latter tradition of Foucauldian STS scholarship in the United States, best represented by scholars like Daston and Galison, that we would like to focus on here as a particularly influential example of what media genealogy might look like.

Galison has explicitly acknowledged the influence that Foucault has had on his work, noting that foremost is Foucault's focus on the material dimension of knowledge production. He notes that one key difference is that he tends to focus on much more tangible technologies than Foucault did.[102] This is clear in Galison's move to extend Foucault's notion of the "conditions of possibility" into the realm of scientific instruments in his account of "conditions of instrumentality."[103] In *The Archaeology of Knowledge*, Foucault attempted to describe how certain statements came to be used rather than others.[104] He asked what conditions needed to be in place to make it possible to describe the world in a particular way and be able to claim the description was true and accurate. Galison extends this sensibility by focusing attention upon a similar co-dependent process. These conditions address (1) how specific instruments (media) need to be present to make particular statements or discursive utterances possible, and (2) that only under certain conditions can particular instruments come into being and/or seem necessary in the first place. Simply, what can and should be made representable? In large part the criteria determining the use of instruments accords with the degree to which such forms of instrumentality produce or reduce "noise." Thus, a communications sensibility orients the "conditions of instrumentality." The necessary fidelity of an instrument is of course historical and flexible, but a statement about its fidelity is determinative of what sorts of discourse are recorded, verified, processed, and disseminated. In other words, there is no overarching schema for determining the acceptability of an instrument's clarity, but a concern with clarity conditions such use and helps configure who uses such instruments. Jonathan Sterne's *The Audible Past: Cultural Origins of Sound Reproduction* is a significant investigation into such "instrumentality" providing, for example, an analysis of "a new medical semiotics" in which "doctors could hear what they could not see."[105] And lastly, as Galison notes, instrumental knowledge production is coupled tightly with power.[106]

Galison wants to move beyond Kuhn's model of wholesale paradigmatic change to a model that is more complex and more nuanced and which pays attention to both the work of translation (a linguistic communicative problem) and decontextualization (how do things get translated *and* newly "instrumentalized"). The reason the term *media instrument* holds such potential value as opposed to *media technology* is that the instrument already implies some effect, some agency, over the very possibilities for representing and processing the world into data. Instruments are not merely facilitators or tools used at the behest of human agency. Further, we might take up the relationship Galison develops between science, instrumentality, and subjectification. In "The Collective Author," Galison looks to the contingency of authorship in physics.[107] There he begins to develop an analysis for the examination of "technologies of a scientific self." Distributed authorship is one technological aspect to a collective form of scientific subjectivity, different from other sciences and practices of authorship. Physics projects modes of subjectification.

More fundamentally, in *Objectivity*, Daston and Galison argue that the formation of what has come to be understood as scientific objectivity is largely the result of how new instruments for producing and representing data are intertwined with what they call "epistemic virtues." They work through three differing forms of such virtue: truth-to-nature, objectivity, and trained judgment. What they all share is a belief that instruments and discursive rules exist which when properly used can overcome subjectivity, with all of its misperceptions and perspectival limitations. Scientific atlases, filled with hand-drawn or mechanically produced images, charts, X-rays, echocardiograms, and the like, have all been used to teach the proper means for seeing and thus knowing the world. As they note, "Once internalized by a scientific collective, these various ways of seeing were lodged deeper than evidence; they defined what evidence was."[108] Hence, an ink-fed needle marking a continuous line that jumped vertically up and down across a series of predetermined horizontal lines became the means for objectively representing, and hence knowing, the relative health of the cardiovascular system. Yet a trained reader, ethically committed to an epistemological framework, needed to be simultaneously produced to integrate these markings into an institutional network that could fix this heart so it could later produce a healthy chiaroscuro of markings that fell within the limits of the norm. This co-dependent relationship between technology and subjectification determines what technological forms get developed for use while simultaneously legitimating an understanding of the world that is fundamentally mediated by those same technologies. In *Histories of Scientific Observation*, Daston and Elizabeth Lunbeck provide a summary of how such processes work together: "Observation educates the senses, calibrates judgement, picks out objects of scientific inquiry, and forges 'thought collectives.' "[109] It has increasingly done so by way of technological instruments *or* media.

So what can we take from this for the study of media history from a Foucauldian perspective? Perhaps most important, in experimental (as well as the theoretical) sciences, the production of discourse is dependent upon the "possibilities of instrumentality." The mediating technologies that respond to and produce observable representations of physical phenomena are integral to an understanding of media, as are the technologies of reproduction and dissemination (Kittler's camera obscura and lanterna magica, for instance). In other words, all knowledge is dependent upon forms of technologically mediated discourse. Whether this runs through the eye (observation) and the hand (a picto-drawing or lab notes) or digital representations resulting from sound waves (3D ultrasound), the process of mediating experience into discourse (statements that make a claim on the world) is a technological process. So any study of discourse needs to attend to what Galison calls "possibilities of instrumentation." We might further suggest that any study of media history should also. The kinds of statements that could be inscribed at any given moment depended upon the forms and types of instruments present, available, and in use. The world has become representable and knowable through media technologies.

With the recent rise of a totalizing, security-maintaining apparatus that we have described as "the circuit," STS's subject/power critique—that is, its cultural studies side—has had a difficult time understanding its own power and subject position. Specifically, a global trend to the political right has contributed to a broadly perceived "attack on science." In its wake, STS projects that identify and problematize power relations can be perceived as (and in some cases even used as) support for the attack. In 2009, Daston famously articulated a widely perceived ebb in enthusiasm for science studies in Europe and asked a markedly genealogical question: "What can these (troubling, deteriorating) developments tell us about disciplinarity—its preconditions, its practices, its ethos?"[110] She goes on to describe a tendency for STS to stay safely in academic discursive fields, describing the minutiae of specific scientific practices in "miniature" contexts. Here we can understand Daston as pressing on STS's version of the problems we've outlined with German media studies: an overcommitment to archaeology at the expense of genealogy. And this matters, as it is only with the shift to genealogical method that historical research becomes a history of the present. Daston ends the essay with a ray of hope emanating specifically from Foucauldian work. The "counterweight to these miniaturizing tendencies," she writes, "has been supplied . . . by a still more thoroughgoing form of historicism, namely, the philosophical history of Michel Foucault."[111]

Toward Media Genealogy

In his first extended work, Kittler described the object of his analysis as an apparatus composed of networks of technologies and institutions that operate at the level of culture by facilitating data collection, processing, storage, and transmission.[112] In our opinion, Kittler's (often implicit) methodologies are still perhaps the most robust and effective tools for performing media studies in terms of both archaeology and genealogy. For Kittler, technical media emerge piecemeal from a historically conditioned discourse network, combining, mixing, cross-pollinating, and eventually crystallizing, all while gaining an increasing autonomy from the milieu in which they originated. A Kittlerian must examine corporations, militaries, bureaucracies, nonprofit organizations, academies, inventors and development teams, and potentially spatially and temporally distributed contributions in the form of tweakings, developments, optimizations, alterations of functions, combinations, ad infinitum. This is perhaps the best example of an investigation that responds to the call to navigate between technological determinism and symptomatic technology, the Scylla and Charybdis of media studies.

Yet media encompass a much broader range of technologies, all of which serve to articulate and link things together in networks of forces, practices, knowledge, and institutions.[113] It was in this sense that even infrastructures were always already media and that media studies was required to take in a wide-ranging set of discursive and nondiscursive utterances, statements, and grammars of architectures, diagrams, and backup plans that all work to hold together a given, and sometimes fragile, apparatus. We would like to expand this definition in response to our analysis of Kittler's work, and articulate media as tools of governance that shape knowledge and produce and sustain power relations while simultaneously forming their attendant subjects. Media technologies are precisely those that allow for the extension of culture across time, for culture's duration and endurance. As such, they have a priori stakes in the realms of the political, the ethical, and the epistemological. Media collect, store, process, and transmit data that are variously used to rate, coordinate, create, obfuscate, obliterate, translate, demonstrate, and even create virtuality, materiality, and reality itself. Yet we can see this rise as immanent to governance as it has taken shape across the globe in unevenly dispersed fits and starts over the past several thousand years.

What was already in embryo in Kittler's work, and has subsequently gone underdeveloped as German media studies turned toward developing a media archaeological method, was an explicit analysis of the visible surface of contesting forces and power relations on which archaeological depth is laid out. In his definition of discourse networks, Kittler was already close to a notion of problematization that could investigate the fundamental role that media technologies play in determining the conditions of possibility, existence, and truth that articulate and define both subjects and objects in a given culture. Media archaeology's interest in the concrete specificity of an individual technology can miss its larger role in the production and maintenance of a larger apparatus, even though that technology's spatial and temporal location in such an apparatus is immanent to that very technology in its concrete specificity.

It is worth reiterating here that our extended engagement with media archaeology's methods is not meant to discourage scholars from continuing to perform archaeological investigations of media and technologies. Instead, we have tried to demonstrate how that methodological commitment leads outside of itself, that at some point it requires a genealogical component, which, when added, lends media studies a relevance and urgency it might not otherwise have. We are also of the opinion that opening media archaeology up to genealogical commitments—notably, power and subjectivation—allows media studies to better interface with hugely significant and often overlapping investigations from other disciplines of media, science, governance, and technology.

Conclusion

This book, then, is, among other things, a set of interrelated cases that can serve to demonstrate the methodological scope of what we hope is a sufficient media genealogy: good archaeology, informed by media theory and media history, and good genealogy, informed by cultural studies. It also attempts the cohesive articulation of a problematic, the circuit, that comprises media-enabled governance of overlapping realms of human existence, including war, medical management and public health, policing, automotive transportation, and human vision. We and many of the scholars we draw from in this book see an emergent mode of such relations, inescapable circuits that exist to gather information and then activate power in ways that are increasingly totalized, simultaneous, instantaneous, and imperceivable. The circuit's apparatus minimizes interference and maximizes efficiency by producing media that eschew mediality via heterogeneous infrastructures with automated processes that include our bodies going through the motions of daily life.

Our explicitly Foucauldian approach is hobbled in the same way that all also-imbricated critique of any form of governance is. The act of interrogating power, especially in an academic context, can seem ironically impotent. We feel keenly the ubiquitous critiques of critique that say there is nothing outside of the critical move, that the theorists who engage it use it to avoid declaring a stance. Though the examination of an apparatus may serve to reveal or interrogate the politics that produced it and the politics that it reproduces, a similar apparatus enables and produces the critique itself. It would be shortsighted and hubristic at best and unthinkingly harmful at worst, then, for us to say that this work, in the form of this book, is a liberatory political act in itself. This book is not activism, though the collaborative spirit in which it was written may well have been.

Nevertheless, the first move of any escape plan or tactical sabotage must be to map out the walls of the prison house. We intend for our work, tying the meticulous technical minutiae of archaeology to the power-interrogating method of genealogy, to further empower a transformative liberatory imaginary. At the end of his 1983 lectures to distill a theoretical history of cultural studies, Stuart Hall writes:

> People have to have a language to speak about where they are and what other possible futures are available to them. These futures may not be real; if you try to concretise them immediately, you may find there is nothing there. But what is there . . . is the possibility of being someone else, of being in some other social space from the one in which you have already been placed.[114]

We see an encircuited foreclosure of that possibility in progress. And we would add that in addition to a language, people have to have the tools to make a different future and the wherewithal not to rebuild the same one.

Notes

Preface

1. See Michel Foucault, *The Archaeology of Knowledge and the Discourse on Language*, trans. A. M. Sheridan Smith (New York: Pantheon, 1972).

An Introduction to the Circuit

1. "Far Worse than Hanging; Kemmler's Death Proves an Awful Spectacle," *New York Times*, August 6, 1890.
2. "Far Worse than Hanging."
3. Mark Essig, *Edison and the Electric Chair: A Story of Light and Death* (New York: Walker and Co., 2005), 20.
4. "Far Worse than Hanging."
5. "Far Worse than Hanging."
6. Jürgen Martschukat, " 'The Art of Killing by Electricity': The Sublime and the Electric Chair," *Journal of American History* 89, no. 3 (2002): 900–921. While we are not arguing that electrocution is humane, Martschukat's work on the electric chair and the sublime explains that the fervor around electricity seemingly promised to make all things better: "Electricity promised to reduce the moment of dying to a split second and to strip death of its supposedly archaic characteristics. In the guise of electrified civilization, death could occur without being associated with struggle, sorrow, and bodily destruction" (911). He explains that this was often not the case, with many of these executions proceeding despite the promise of better death.
7. Tom McNichol, *AC/DC: The Savage Tale of the First Standards War* (San Francisco: Wiley, 2006).
8. "Durston (the prison warden) was pacing the room, angry that the witnesses were late in arriving. He needed the execution to be over by seven, so that the steam engines being used to power the dynamos could begin their usual work of running machinery in the prison's factories" (Essig, *Edison and the Electric Chair,* 20).
9. Essig, 20.
10. These patients were labeled with the acronym YAVIS by the mental health community.

11. Accounts from the United States and Britain report that electrodes were also attached to the testicles in some therapy settings.

12. Douglas C. Haldeman, "Sexual Orientation Conversion Therapy for Gay Men and Lesbians: A Scientific Examination," in *Homosexuality: Research Implications for Public Policy*, ed. John C. Gonsiorek and James D. Weinrich (Newbury Park, Calif.: Sage, 1991), 160.

13. Such therapies were later proven entirely ineffective and, more important, unethical and inhumane. See Haldeman.

14. Dieter Bohn, "Elon Musk: Negative Media Coverage of Autonomous Vehicles Could Be 'Killing People,'" *The Verge*, October 19, 2016, https://www.theverge.com/2016/10/19/13341306/elon-musk-negative-media-autonomous-vehicles-killing-people.

15. "Circuit," *Oxford Dictionary of English*, 316.

16. "Circuit."

17. John Durham Peters, "Calendar, Clock, Tower," in *Deus in Machina: Religion, Technology, and the Things in Between*, ed. Jeremy Stolow (New York: Fordham University Press, 2013).

18. Michel Foucault, *The Foucault Effect: Studies in Governmentality* (Chicago: University of Chicago Press, 1991), 93.

19. Matthew Tiessen and Greg Elmer, "Editorial Introduction: Deleuze/Foucault: A Neoliberal Diagram," *MediaTropes* 4, no. 1 (2013): i.

20. Mark Usher, "Veins of Concrete, Cities of Flow: Reasserting the Centrality of Circulation in Foucault's Analytics of Government," *Mobilities* 9, no. 4 (October 2014): 550–69.

21. Usher, 551.

22. Claudia Aradau and Tobias Blanke, "Governing Circulation: A Critique of the Biopolitics of Security," in *Security and Global Governmentality: Globalization, Governance, and the State*, ed. Miguel de Larrinaga and Marc G. Doucet (Abingdon, UK: Routledge, 2010).

23. Michel Foucault, *Psychiatric Power: Lectures at the Collège de France, 1973–1974* (New York: Macmillan, 2008).

24. Michel Foucault, *Security, Territory, Population: Lectures at the Collège de France, 1977–1978*, ed. Michel Senellart, trans. Graham Burchell (New York: Picador, 1977).

25. Michel Foucault, *Discipline and Punish: The Birth of the Prison* (New York: Pantheon, 1977), 29.

26. Foucault, Ewald, and Fontana, *Security, Territory, Population*, 25.

27. Mark B. Salter, *Making Things International 1: Circuits and Motion* (Minneapolis: University of Minnesota Press, 2015).

28. Sean Cubitt, *The Practice of Light: A Genealogy of Visual Technologies from Prints to Pixels* (Boston: MIT Press, 2014).

29. Chris Philo argues that it is in Foucault's most unlikely texts that we find a concern with the necessary presence of circulation as in for instance the asylum. See Philo, "'One Must Eliminate the Effects of . . . Diffuse Circulation [and] Their Unstable and Dangerous Coagulation': Foucault and Beyond the Stopping of Mobilities," *Mobilities* 9, no. 4 (October 2014): 493–511.

30. Paul Virilio, in *Speed and Politics* (Boston: MIT Press, 1986), suggests such an understanding. Packer, in *Mobility without Mayhem* (Durham, N.C.: Duke University Press, 2008), argues that "disciplined mobility" is a governmental priority of automobility, not an oxymoron.

31. Michel Foucault, *Power: Essential Works of Michel Foucault, 1954–1984* (New York: New Press, 2001), 364.

32. Paul Rabinow, *The Foucault Reader* (New York: Pantheon Books, 1984), 256.

33. Michael C. Behrent, "Foucault and Technology," *History and Technology* 29, no. 1 (2013): 54–104.

34. Friedrich Kittler *Gramophone, Film, Typewriter* (Stanford: Stanford University Press, 1999), xl–xli.

35. Stefan Elbe, "Our Epidemiological Footprint: The Circulation of Avian Flu, SARS, and HIV/AIDS in the World Economy," *Review of International Political Economy* 15, no. 1 (2007): 116–30.

36. Peter J. Forman, "Circulations beyond Nodes: (In)securities along the Pipeline," *Mobilities* 13, no. 2 (March 2018): 231–45.

37. Paul N. Edwards, *The Closed World: Computers and the Politics of Discourse in Cold War America* (Boston: MIT Press, 1996).

38. Gilles Deleuze and Félix Guattari, *A Thousand Plateaus: Capitalism and Schizophrenia*, trans. Brian Massumi (Minneapolis: University of Minnesota Press, 1987).

39. Bernhard Siegert, "The Map Is the Territory," *Radical Philosophy* 5 (2011): 13–16.

40. Armand Mattelart, *The Invention of Communication* (Minneapolis: University of Minnesota Press, 1996).

41. Mattelart, xiv.

42. See Evelyn Fox Keller, *Refiguring Life: Metaphors of Twentieth-Century Biology* (New York: Columbia University Press, 1995), for machinic genetic metaphors; see George Lakoff and Mark Johnson, *Metaphors We Live By* (Chicago: University of Chicago Press, 2008), for computational metaphors; and see N. Katherine Hayles, *My Mother Was a Computer: Digital Subjects and Literary Texts* (Chicago: University of Chicago Press, 2010), for computational worldviews.

43. Foucault, Ewald, and Fontana, *Security, Territory, Population, 18.*

44. Foucault, Ewald, and Fontana, 18.

45. Foucault, Ewald, and Fontana, 19.

46. Friedrich Kittler, "There Is No Software," *Ctheory* (1995): 10–18.

47. Aden Evens, *Logic of the Digital* (London: Bloomsbury Publishing, 2015); Lev Manovich, *The Language of New Media* (Boston: MIT Press, 2002); Eugene Thacker, *Biomedia* (Minneapolis: University of Minnesota Press, 2004); N. Katherine Hayles, *How We Think: Digital Media and Contemporary Technogenesis* (Chicago: University of Chicago Press, 2012).

48. See especially Kittler's archaeological projects: *Gramophone, Film, Typewriter;* and *Discourse Networks, 1800/1900* (Stanford: Stanford University Press, 1990).

49. See James Gleick, *The Information: A History, a Theory, a Flood* (New York: Vintage, 2011); George Dyson, *Turing's Cathedral: The Origins of the Digital Universe* (New York: Vintage, 2012).

50. Benjamin H. Bratton, *The Stack: On Software and Sovereignty* (Boston: MIT Press, 2016), 33.

51. Joshua Reeves, *Citizen Spies: The Long Rise of America's Surveillance Society* (New York: NYU Press, 2017).

52. Colin Koopman, *How We Became Our Data: A Genealogy of the Informational Person* (Chicago: University of Chicago Press, 2019).

53. Aden Evens, "Web 2.0 and the Ontology of the Digital," *Digital Humanities Quarterly* 6, no. 2 (2012).

54. Ted Striphas, "Culture," in *Digital Keywords: A Vocabulary of Information Society and Culture*, vol. 8, ed. Benjamin Peters (Princeton: Princeton University Press, 2016), 70–80.

55. Hayles, *My Mother Was a Computer.*

56. Yuk Hui, *On the Existence of Digital Objects* (Minneapolis: University of Minnesota Press, 2016).

57. W. H. Auden, "Oxford," in *Collected Shorter Poems, 1930–1944* (London: Faber & Faber, 1950), 98.

58. John von Neumann, "Probabilistic Logics and the Synthesis of Reliable Organisms from Unreliable Components," *Automata Studies* 34 (1956): 43–98.

59. Paul Baran, *Reliable Digital Communications Systems Utilizing Unreliable Network Repeater Nodes* (RAND Corporation Memorandum P-1995, May 27, 1960).

60. Donald Davies, *Proposal for a Digital Communication Network* (National Physical Laboratory, June 1966).

61. George B. Dyson, *Darwin among the Machines: The Evolution of Global Intelligence* (Reading, Mass.: Helix Books, 1997), 205.

62. Alexander R. Galloway and Eugene Thacker, *The Exploit: A Theory of Networks* (Minneapolis: University of Minnesota Press, 2007), 5.

63. Galloway and Thacker, 32.

64. J. C. R. Licklider and Robert W. Taylor, "The Computer as a Communication Device," *Science and Technology* 76, no. 2 (1968): 1–3.

65. Yuk Hui, "Modulation after Control," *New Formations: A Journal of Culture/Theory/Politics* 84–85 (2015): 74–91.

66. Alexander R. Galloway, *Protocol: How Control Exists after Decentralization* (Cambridge: MIT Press, 2004), 3.

67. Gilles Deleuze, "Postscript on the Societies of Control," *October* 59 (Winter 1992): 4.

68. Deleuze, 5.

69. Galloway and Thacker, *The Exploit,* 5.

70. Antoinette Rouvroy, "The End(s) of Critique: Data Behaviourism versus Due Process," in *Privacy, Due Process, and the Computational Turn: Philosophers of Law Meet Philosophers of Technology*, ed. Mireille Hildebrandt and Ekatarina de Vries (New York: Routledge, 2012), 143–68. See also Tiziana Terranova, *Network Culture: Politics for the Information Age* (London: Pluto, 2004), 56; and John Cheney-Lippold, *We Are Data: Algorithms and the Making of Our Digital Selves* (New York: NYU Press, 2017), 107.

71. Deleuze, "Postscript on the Societies of Control," 3–7.

72. Hui, "Modulation after Control," 75.

73. John-Christophe Plantin, Carl Lagoze, Paul N. Edwards, and Christian Sandvig, "Infrastructure Studies Meet Platform Studies in the Age of Google and Facebook," *New Media & Society* 20, no. 1 (2018): 293–310.

74. Wiebe Bijker, Thomas P. Hughes, and Trevor Pinch, *The Social Construction of Technological Systems* (Cambridge: MIT Press, 1987); Thomas P. Hughes, *Networks of Power: Electrification in Western Society, 1880–1930* (Baltimore: Johns Hopkins University Press, 1988); Renate Mayntz and Thomas P. Hughes, *The Development of Large Technical Systems* (Boulder: Westview Press, 1988); Nicole Starosielski, *The Undersea Network* (Durham, N.C.: Duke University Press, 2015).

75. Paul N. Edwards, "Infrastructure and Modernity: Force, Time, and Social Organization in the History of Sociotechnical Systems," *Modernity and Technology* 1 (2003): 185.

76. Geoffrey C. Bowker and Susan Leigh Star, *Sorting Things Out: Classification and Its Consequences* (Cambridge: MIT Press, 1999); Paul N. Edwards, Geoffrey C. Bowker, Steven J. Jackson, and Robin Williams, "Introduction: An Agenda for Infrastructure Studies," *Journal of the Association for Information Systems* 10, no. 5 (2009): 364–74; David Ribes and Thomas Finholt, "The Long Now of Technology Infrastructure: Articulating Tensions in Development," *Journal of the Association for Information Systems* 10 (2009): 375–98.

77. Tarleton Gillespie, *Custodians of the Internet: Platforms, Content Moderation, and the Hidden Decisions That Shape Social Media* (New Haven, Conn.: Yale University Press, 2018), 18, 21.

78. José van Dijck, *The Culture of Connectivity: A Critical History of Social Media* (New York: Oxford University Press, 2013), 26.

79. Ganaele Langlois and Greg Elmer, "The Research Politics of Social Media Platforms," *Culture Machine* 14 (2013), available at http://www.culturemachine.net/index.php/cm/article/view/505.

80. Carliss Y. Baldwin and C. Jason Woodward, "The Architecture of Platforms: A Unified View," in *Platforms, Markets, and Innovation*, ed. Annabelle Gawer (Northampton, Mass.: Edward Elgar Publishing, 2011), 32.

81. Plantin et al., "Infrastructure Studies Meet Platform Studies."

82. Tarleton Gillespie, "The Politics of 'Platforms,'" *New Media & Society* 12, no. 3 (2010): 347–64.

83. Nick Srnicek, *Platform Capitalism* (Brooklyn: Polity, 2017), 48.

84. Srnicek, 113.

85. Jonathan Crary, *Techniques of the Observer* (Cambridge, Mass.: MIT Press, 1990); Lorraine Daston and Peter Galison, *Objectivity* (Princeton, N.J.: Princeton University Press, 2021).

86. Friedrich Kittler, *Optical Media* (Cambridge, UK: Polity, 2010); and Cubitt, *The Practice of Light.*

87. Couze Venn, Roy Boyne, John W. P. Phillips, and Ryan Bishop. "Technics, Media, Teleology: Interview with Bernard Stiegler." *Theory, Culture & Society* 24, nos. 7–8 (2007): 339.

88. Venn et al., 338.

89. Michel Foucault, *Power/Knowledge: Selected Interviews and Other Writings, 1972–1977* (New York: Vintage, 1980).

90. *A Clockwork Orange*, directed by Stanley Kubrick (Warner Bros. Pictures, 1971).

91. *Battlestar Galactica*, season 2, episode 3, "Fragged," directed by Sergio Mimica-Gezzan, aired July 29, 2005, on SciFi Channel.

92. A trope perennially used to respond rhetorically to "risky" technoscientific endeavors; see Sheryl N. Hamilton, "Traces of the Future: Biotechnology, Science Fiction, and the Media." *Science Fiction Studies* 30, no. 2 (2003): 267–82.

1. How to Make a Soldier into a Medium

1. Friedrich Kittler, *Discourse Networks 1800/1900*, trans. Michael Metteer (Stanford: Stanford University Press, 1990).

2. N. Katherine Hayles, *How We Think: Digital Media and Contemporary Technogenesis* (Chicago: University of Chicago Press, 2012).

3. Sybille Krämer, *Medium, Messenger, Transmission: An Approach to Media Philosophy* (Amsterdam: Amsterdam University Press, 2015).

4. The Signal Corps was founded at a crucial moment in the development of communications infrastructure (at the same time as the telegraph), and the Corps could arguably be said to have created the first Communications Depart-

ment in U.S. higher education: Albert Myer's school for signal instruction at Red Hill in Georgetown, Washington, D.C.

5. Krämer, *Medium, Messenger, Transmission.*

6. N. Katherine Hayles, *My Mother Was a Computer: Digital Subjects and Literary Texts* (Chicago: University of Chicago Press, 2005); Lev Manovich, *The Language of New Media* (Cambridge, Mass.: MIT Press, 2002); Aden Evens, "Web 2.0 and the Ontology of the Digital," *Digital Humanities Quarterly* 6, no. 2 (2012).

7. James W. Carey, *Communication as Culture: Essays on Media and Society*, rev. ed. (New York: Routledge, 1989); Kittler, *Discourse Networks*; John Durham Peters, "Technology and Ideology: The Case of the Telegraph Revisited," in *Thinking with James Carey: Essays on Communications, Transportation, History*, ed. Jeremy Packer and Craig Robertson (New York: Peter Lang, 2006), 137–55; Hayles, *How We Think*; David Hochfelder, *The Telegraph in America, 1832–1920* (Baltimore: Johns Hopkins University Press, 2012).

8. Friedrich Kittler. "The History of Communication Media," *Ctheory* (1996): 7–30.

9. Carey, *Communication as Culture*, 32.

10. Sybille Krämer, "The Cultural Techniques of Time Axis Manipulation: On Friedrich Kittler's Conception of Media," *Theory, Culture & Society* 23, nos. 7–8 (2006): 96.

11. James W. Carey, "Historical Pragmatism and the Internet," *New Media & Society* 7, no. 4 (2005): 443–55.

12. Geoffrey Winthrop-Young, *Kittler and the Media* (Cambridge, UK: Polity, 2011), 137.

13. Friedrich A. Kittler, "Media Wars: Trenches, Lightning, Stars," in *Literature, Media, Information Systems*, ed. John Johnston (Abingdon, UK: Routledge, 1997), 119.

14. Peters, "Technology and Ideology," 138.

15. Hayles, *How We Think*, 127.

16. Hayles, 147.

17. David Bates, *Lincoln in the Telegraph Office: Recollections of the United States Military Telegraph Corps during the Civil War* (New York: Century Co., 1907).

18. Hochfelder, *The Telegraph in America*, 11.

19. Albert Myer, "A New Sign Language for Deaf Mutes," *Buffalo Medical Journal and Monthly Review of Medical and Surgical Science* 7, no. 1 (1851): 14–20.

20. Kittler, *Discourse Networks*, 23.

21. Kittler, 22.

22. Kittler, 370; Mara Mills and Jonathan Sterne, "Afterword II: Dismediation: Three Proposals, Six Tactics," in *Disability Media Studies*, ed. Elizabeth Ellcessor and Bill Kirkpatrick (New York: NYU Press, 2017), 365–78.

23. Albert James Myer, *A Manual of Signals: for the Use of Signal Officers in the Field, and for Military and Naval Students, Military Schools, Etc: A New Ed., Enlarged and Illustrated* (New York: D. van Nostrand, 1868), 56.

24. Kittler, *Discourse Networks*, 2.

25. *A History of the U.S. Signal Corps* (New York: Putnam, 1961).

26. *A History of the U.S. Signal Corps*, 21.

27. Michel Foucault, *Discipline and Punish: The Birth of the Prison*, trans. Alan Sheridan (New York: Random House Digital, Inc., 1977), 47.

28. Harry Davis and Frederick G. Fasset, *What You Should Know about the Signal Corps* (New York: Norton, 1943), 137.

29. *A History of the U.S. Signal Corps*, 76.

30. Hochfelder, *The Telegraph in America*, 11.

31. *A History of the U.S. Signal Corps*, 11.

32. U.S. Signal Corps, *Property and General Regulations of the Signal Corps* (Washington, D.C.: U.S. Army Signal Corps, 1898), 76.

33. *A History of the U.S. Signal Corps*, 14.

34. Rebecca Robbins Raines, *Getting the Message Through: A Branch History of the U.S. Army Signal Corps* (Honolulu: University Press of the Pacific, 2005), 3.

35. Kittler. "The History of Communication Media."

36. Jeremy Packer, "Screens in the Sky: SAGE, Surveillance, and the Automation of Perceptual, Mnemonic, and Epistemological Labor," *Social Semiotics* 23, no. 2 (2013): 173–95.

37. Robbins Raines, *Getting the Message Through*, 37.

38. J. Willard Brown, *The Signal Corps, U.S.A., in the War of the Rebellion* (Boston: U.S. Veteran Signal Corps Association, 1896), 104.

39. Brown, 106–7.

40. Friedrich Kittler, *Gramophone, Film, Typewriter*, trans. Geoffrey Winthrop-Young and Michael Wutz (Stanford: Stanford University Press, 1999), 191.

41. Robbins Raines, *Getting the Message Through*, 16.

42. Robbins Raines, 16.

43. Robbins Raines, 27.

44. *A History of the U.S. Signal Corps*, 15.

45. Greg Elmer, "A New Medium Goes Public: The Financialization of Marconi's Wireless Telegraph & Signal Company," *New Media & Society* 19, no. 11 (2016): 1829–47.

46. Norman Ohler, *Blitzed: Drugs in the Third Reich*, trans. Shaun Whiteside (Boston: Houghton Mifflin Harcourt, 2017).

47. Robbins Raines, *Getting the Message Through*, 16.

48. *A History of the U.S. Signal Corps*, 24.

49. *A History of the U.S. Signal Corps*, 27.

50. Hochfelder, *The Telegraph in America*, 12.

51. Kittler, *Gramophone, Film, Typewriter*, 249.

52. George Dyson, *Turing's Cathedral: The Origins of the Digital Universe* (New York: Pantheon, 2012).

53. Advertisement, *Popular Science* 141, no. 4 (1942): 2.

54. Davis and Fasset, *What You Should Know*, 42.

55. Davis and Fasset, 42.

56. Dyson, *Turing's Cathedral.*

57. Evens, "Web 2.0," 6.

58. Myer, "A New Sign Language."

59. Cixin Liu, *The Three-Body Problem*, trans. Ken Liu (New York: Tor, 2006).

60. Michel Foucault, *The History of Sexuality, Volume 1: An Introduction*, trans. Robert Hurley (New York: Vintage, 1978).

61. Michel Foucault, *Madness and Civilization: A History of Insanity in the Age of Reason*, trans. Richard Howard (New York: Vintage, 1988), xii.

62. See Joshua Reeves, "Of Social Networks and Suicide Nets: Biopolitics and the Suicide Screen," *Television & New Media* 18, no. 6 (2017): 512–28.

2. Soldiers in the Circuit

1. Friedrich Kittler, *Literature, Media, Information Systems* (New York: Routledge, 1997), 73.

2. John Durham Peters, "'The Only Proper Scale of Representation': The Politics of Statistics and Stories," *Political Communication* 18, no. 4 (2001): 436, 435.

3. See Craig Robertson, *The Passport in America: The History of a Document* (New York: Oxford University Press, 2010).

4. David T. Mitchell and Sharon L. Snyder, eds., *The Body and Physical Difference: Discourses of Disability* (Ann Arbor: University of Michigan Press, 1997), 8.

5. Mara Mills and Jonathan Sterne, "Afterword II: Dismediation—Three Proposals, Six Tactics," in *Disability Media Studies*, ed. Elixabeth Ellcessor and Bill Kirkpatrick (New York: NYU Press, 2017), 365–78.

6. Colin Koopman, "Information before Information Theory: The Politics of Data beyond the Perspective of Communication," *New Media & Society* 21, no. 6 (2019): 1326–43; Alexander Monea and Jeremy Packer, "Media Genealogy and the Politics of Archaeology," *International Journal of Communication* 10 (2016): 3141–59; J. J. Sylvia IV, "From Archaeology to Genealogy: Adding Processes of Subjectivation and Artistic Intervention," *communication +1* 1, no. 2 (2019): 1–25.

7. For example, Paul Ceruzzi, *A History of Modern Computing*, 2nd ed. (Cambridge: MIT Press, 2003); Paul Edwards, *The Closed World: Computers and the Politics of Discourse in Cold War America* (Cambridge: MIT Press, 1996);

Friedrich Kittler, *Gramophone, Film, Typewriter,* trans. Geoffrey Winthrop-Young and Michael Wutz (Stanford: Stanford University Press, 1999).

8. Geoffrey Winthrop-Young, "Drill and Distraction in the Yellow Submarine: On the Dominance of War in Friedrich Kittler's Media Theory," *Critical Inquiry* 28, no. 4 (2002): 845.

9. Winthrop-Young, 845–46.

10. Winthrop-Young, 845.

11. Paul Virilio, *Speed and Politics: An Essay on Dromology,* trans. Mark Polizzotti (New York: Semiotext(e), 1986); Paul Virilio, *War and Cinema: The Logistics of Perception,* trans. Patrick Camillier (New York: Verso, 1989).

12. Keith Hoskin, "Education and the Genesis of Disciplinarity: The Unexpected Reversal," in *Knowledges: Historical and Critical Studies in Disciplinarity,* ed. Ellen Messer-Davidow and David Sylvan (Charlottesville: University Press of Virginia, 1993), 271–304; Keith Hoskin and Richard Macve, *Power through Knowledge: The Logic of Modern Business* (New York: Oxford University Press, 2009).

13. James Beniger, *The Control Revolution: Technological and Economic Origins of the Information Society* (Cambridge, Mass.: Harvard University Press, 1989).

14. Nathan Ensmenger, *The Computer Boys Take Over: Computers, Programmers, and the Politics of Technical Expertise* (Cambridge: MIT Press, 2012); Jennifer S. Light, "When Computers Were Women," *Technology and Culture* 40, no. 3 (1999): 455–83.

15. Benjamin Gould, *Investigations in the Military and Anthropological Statistics of American Soldiers,* 2 vols. (Cambridge, Mass.: Riverside Press, 1869).

16. "Confederate Widow, 93, Dies in Ark.," *Washington Times,* August 18, 2008, https://www.washingtontimes.com/news/2008/aug/18/maudie-cecilia-hopkins-93-confederate-widow-dies/.

17. Geoffrey Austrian, *Herman Hollerith: Forgotten Giant of Information Processing* (New York: Columbia University Press, 1982), 2.

18. Lars Heide, *Punched-Card Systems and the Early Information Explosion: 1880–1945* (Baltimore: Johns Hopkins University Press, 2009).

19. Austrian, *Herman Hollerith,* 4, 5–7.

20. James Essinger, *Jacquard's Web: How a Hand Loom Led to the Birth of the Information Age* (New York: Oxford University Press, 2004), 37.

21. Austrian, *Herman Hollerith,* 16.

22. Essinger, *Jacquard's Web,* 159.

23. Leon Edgar Truesdell, *The Development of Punch Card Tabulation in the Bureau of the Census, 1890–1940, with Outlines of Actual Tabulation Programs* (Washington, D.C.: U.S. Government Printing Office, 1965), 30–31.

24. Austrian, *Herman Hollerith,* 17.

25. Austrian, 8.

26. Heide, *Punched-Card Systems,* 29.

27. Virginia Hollerith, "Biographical Sketch of Herman Hollerith," *Isis* 62, no. 1 (1971): 72.

28. Herman Hollerith, *U.S. Patent No. 395,782* (Washington, D.C.: U.S. Patent and Trade Office, 1889).

29. Truesdell, *The Development of Punch Card Tabulation,* 39–44.

30. Heide, *Punched-Card Systems,* 47.

31. Heide, 25; Truesdell, *The Development of Punch Card Tabulation,* 47–51.

32. Truesdell, *The Development of Punch Card Tabulation,* 51–52.

33. Heide, *Punched-Card Systems,* 60.

34. Heide, 60–61.

35. Truesdell, *The Development of Punch Card Tabulation,* 53–54.

36. Heide, *Punched-Card Systems,* 31.

37. Heide, 47; Truesdell, *The Development of Punch Card Tabulation,* 51, 55.

38. Heide, *Punched-Card Systems,* 30.

39. Hollerith, "Biographical Sketch of Herman Hollerith," 71–72.

40. Truesdell, *The Development of Punch Card Tabulation,* 56.

41. Enoch Herbert Crowder, *The Spirit of Selective Service* (New York: Century Co., 1920), 261.

42. *Selective Service Regulations: Prescribed by the President under the Authority Vested in Him by the Terms of the Selective Service Law (Act of Congress Approved May 18, 1917)* (Washington, D.C.: U.S. Government Printing Office, 1917), iii.

43. Edwin Black, *IBM and the Holocaust: The Strategic Alliance between Nazi Germany and America's Most Powerful Corporation* (New York: Crown, 2001).

44. *Selective Service Regulations,* iii.

45. *Selective Service Regulations,* 34.

46. *Selective Service Regulations,* 219–20.

47. *Selective Service Regulations,* 192–96.

48. *Selective Service Regulations,* 227–30.

49. *Selective Service Regulations,* 228.

50. *Selective Service Regulations,* 229.

51. Karl M. Bowman, "The Relation of Defective Mental and Nervous States to Military Efficiency," *Military Surgeon* 46 (1920): 651–69.

52. Bowman, 652.

53. Bowman, 669.

54. Bowman, 653.

55. Bowman, 654.

56. Bowman, 654.

57. Stephen Jay Gould, *The Mismeasure of Man,* rev. and exp. ed. (New York: Norton, 1996), 223–24.

58. Clarence Yoakum and Robert Yerkes, *Army Mental Tests* (New York: Henry Holt & Co., 1920), 2.

59. Robert Yerkes, "Report of the Psychology Committee of the National Research Council," *Psychological Review* 26, no. 2 (1919): 83.

60. S. J. Gould, *The Mismeasure of Man*, 240–42.

61. S. J. Gould, 240–42.

62. Vivian A. C. Henmon, "Air Service Tests of Aptitude for Flying," *Journal of Applied Psychology* 3, no. 2 (1919): 103.

63. Jefferson M. Koonce, "A Brief History of Aviation Psychology," *Human Factors* 26, no. 5 (1984): 500.

64. Henmon, "Air Service Tests of Aptitude for Flying," *Journal of Applied Psychology*, 3, no. 2 (1919): 103–9.

65. Crowder, *The Spirit of Selective Service*, 125–26.

66. Crowder, 355–56.

67. Crowder, 318.

68. Kai Tiaki, "The Care of the Wounded," *Journal of the Nurses of New Zealand* 8, no. 2 (1915): 79–80.

69. "How War Wounded Fare: Elaborate System Followed in Caring for the Disabled," *New York Times*, May 29, 1915, https://www.nytimes.com/1915/05/29/archives/how-war-wounded-fare-elaborate-system-followed-in-caring-for-the.html.

70. "An Army Hospital: The Curse of the Gas, A Journey of Mercy at the Dressing Station, Swift and Unerring Sufferings of the Victims Down to the Base," *Times of India*, July 2, 1915.

71. See Martin R. Howard, *Napoleon's Doctors: The Medical Services of the Grande Armée* (Gloucestershire, UK: Spellmount, 2006), chapter 2, "The Battlefield: The Birth of the Ambulance."

72. Howard, 83.

73. George Thompson, "Battlefield Medicine: Triage-Field Hospital Section," *Medicine in the First World War*, http://www.kumc.edu/wwi/military-medical-operations/triage-field-hospital-section.html.

74. Albert Love, Eugene Hamilton, and Ida Hellman, *Tabulating Equipment and Army Medical Statistics* (Washington, D.C.: Office of the Surgeon General Department of the Army, 1958), 53.

75. Lisa Gitelman, *Paper Knowledge: Toward a Media History of Documents* (Durham, N.C.: Duke University Press, 2014).

76. Love, Hamilton, and Hellman, *Tabulating Equipment*, 55.

77. George Sterling Ryerson, *The Soldier and the Surgeon: A Paper Read before the Canadian Military Institute, Toronto, March 6th, 1899* (Toronto: William Briggs, 1899).

78. Ryerson, 3.

79. Ryerson, 4.
80. Ryerson, 5.
81. Ryerson, 8, emphasis added.
82. Ryerson, 8.
83. Neil Gerlach, Sheryl Hamilton, Rebecca Sullivan, and Priscilla Walton, *Becoming Biosubjects: Bodies, Systems, Technology* (Toronto: University of Toronto Press, 2011), 23.
84. Ryerson, *The Soldier and the Surgeon,* 9.
85. Ryerson, 10.
86. Ryerson, 10.
87. William Grant Macpherson, Anthony Albert Bowlby, and Cuthbert Wallace, eds., *Medical Services Surgery of the War,* vol. 1 (London: His Majesty's Stationery Office, 1922).
88. Macpherson, Bowlby, and Wallace, 209.
89. Love, Hamilton, and Hellman, *Tabulating Equipment,* v.
90. Woodrow Wilson, "President's Proclamation of August 31st, 1918, Calling for the Third Registration," in *Second Report of the Provost Marshal General to the Secretary of War on the Operations of the Selective Service System to December 20, 1918* (Washington, D.C.: U.S. Government Printing Office, 1918), 311.
91. Jill Frahm, "The Hello Girls: Women Telephone Operators with the American Expeditionary Forces during World War I," *Journal of the Gilded Age and Progressive Era* 3, no. 3 (2004): 271–93.
92. *Report of the Surgeon General, U.S. Army, to the Secretary of War* (Washington, D.C.: U.S. Government Printing Office, 1916); *Report of the Surgeon General, U.S. Army, to the Secretary of War* (Washington, D.C.: U.S. Government Printing Office, 1918); *Report of the Surgeon General, U.S. Army, to the Secretary of War,* 2 vols. (Washington, D.C.: U.S. Government Printing Office, 1919).
93. Charles Lynch, ed., *The Medical Department of the United States Army in the World War,* 15 vols. (Washington, D.C.: U.S. Government Printing Office, 1921–29).
94. Charles Davenport and Albert Love, *The Medical Department of the United States Army in the World War (Volume XV): Statistics, Part One: Army Anthropology* (Washington, D.C.: U.S. Government Printing Office, 1921).
95. Davenport and Love, 4.
96. Davenport and Love, 40.
97. See also Albert Love and Charles Davenport, *Defects Found in Drafted Men* (Washington, D.C.: U.S. Government Printing Office, 1920).
98. Chris Anderson, "The End of Theory: The Data Deluge Makes the Scientific Method Obsolete," *Wired,* June 23, 2008, https://www.wired.com/2008/06/pb-theory/.
99. Love and Davenport, 333.

100. Gilles Deleuze, "Postscript on the Societies of Control," *October* 59 (1992): 3–7.

101. Crowder, *The Spirit of Selective Service*, 272.

102. Pablo Garcia et al., "Trauma Pod: A Semi-automated Telerobotic Surgical System," *International Journal of Medical Robotics and Computer Assisted Surgery* 5, no. 2 (2009): 136.

103. Andrew Yoo, Gary Gilbert, and Timothy Broderick, "Military Robotic Combat Casualty Extraction and Care," in *Surgical Robotics: Systems Applications and Visions,* ed. Jacob Rosen, Blake Hannaford, and Richard M. Satava (New York: Springer, 2011), 13.

104. Yoo, Gilbert, and Broderick, 14.

105. Pablo Garcia, "Telemedicine for the Battlefield: Present and Future Technologies," in Rosen, Hannaford, and Satava, *Surgical Robotics,* 67.

106. Gopal Sarma, "Electrical Prescriptions (ElectRx)," https://www.darpa.mil/program/electrical-prescriptions.

107. All quotations in this paragraph are from "DARPA and the Brain Initiative," n.d., https://www.darpa.mil/program/our-research/darpa-and-the-brain-initiative.

108. "Giving the Gift of Independence on Fourth of July: Veterans Receive DARPA's LUKE Arm," June 30, 2017, https://www.darpa.mil/news-events/2017-06-30.

3. Police Circuits

1. Itamar Eichner, "Mossad Helped Thwart Major Iranian Terror Attack in France," July 19, 2018, https://www.ynetnews.com/articles/0,7340,L-5313291,00.html; Reuters Staff, "Israeli Cyber Firm NSO in Talks to Buy Fifth Dimension: Reports," November 12, 2018, https://www.reuters.com/article/us-fifth-dimension-m-a-nso/israeli-cyber-firm-nso-in-talks-to-buy-fifth-dimension-reports-idUSKCN1NH0TY.

2. Cheyenne MacDonald, "Japanese Startup Unveils Minority Report–Style AI That Can Spot Shoplifters BEFORE They Steal," *Daily Mail,* March 5, 2019, https://www.dailymail.co.uk/sciencetech/article-6774313/Japanese-startup-unveils-Minority-Report-style-AI-spot-shoplifters-steal.html.

3. Chris Bousquet, "Mining Social Media Data for Policing, the Ethical Way," April 27, 2018, https://www.govtech.com/public-safety/Mining-Social-Media-Data-for-Policing-the-Ethical-Way.html.

4. Peter K. Manning, *Symbolic Communication: Signifying Calls and the Police Response* (Cambridge: MIT Press, 1988), 3.

5. The whistle began to appear in English policing in about 1880, when it succeeded the police rattle. See T. A. Critchley, *A History of Police in England*

and Wales, 900–1966 (London: Constable Press, 1967), 151. Perhaps the earliest logistical medium in English policing history is the horn, which had been introduced into policing patrols at least by 1302. Anne Rieber DeWindt and Edwin Brezette DeWindt argue that this is among the earliest uses of the horn for policing purposes. See DeWindt and DeWindt, *Ramsey: The Lives of an English Fenland Town, 1200–1600* (Washington, D.C.: Catholic University of America Press, 2006), 331n66. Noting that police patrols at this time were civilian-run and organized around pacts of mutual responsibility, police official and legal historian Frederick Pollock implies that households were required to keep, along with knives and bows, horns to help in policing efforts. See Pollock, *The History of English Law before the Time of King Edward,* vol. 2 (London: C. J. Clay and Sons, 1907), 577. In the fifteenth century, night watchmen—who performed police duties at night—were required to carry horns as they patrolled their villages. These watchmen—who were eventually outfitted with bells and trumpets—were moved to carefully distributed police watchtowers, from which they would send customized alerts based upon the threat to the community. See Thomas Dudley Fosbroke, *Encyclopedia of Antiquities, and Elements of Archaeology, Ancient and Medieval,* vol. 1 (London: John Nichols and Son, 1825), 472.

6. See R. W. Stewart, "The Police Signal Box: A 100-Year History," *Engineering Science Education Journal* 3, no. 4 (1994): 161–68.

7. Rachel Hall, *Wanted: The Outlaw in American Visual Culture* (Charlottesville: University of Virginia Press, 2009).

8. Kathleen Battles, *Calling All Cars: Radio Dragnets and the Technology of Policing* (Minneapolis: University of Minnesota Press, 2010).

9. See, for instance, John Durham Peters, "Calendar, Clock, Tower," in *Deus in Machina: Religion, Technology, and the Things in Between,* ed. Jeremy Stolow (New York: Fordham University Press, 2012), 25–42, in which he argues the logistical and organizational roles of media have been far too often overlooked.

10. Peter Miller and Nikolas Rose, "Governing Economic Life," *Economy and Society* 19 (1990): 1–31. Miller and Rose borrow this framework from Ian Hacking's influential *Representing and Intervening: Introductory Topics in the Philosophy of Natural Science* (New York: Cambridge University Press, 1983).

11. Miller and Rose, 7–8.

12. Here we borrow from Miller and Rose, who synthesized Foucault and Latour to explore "government at a distance."

13. Kelly Gates, "The Tampa 'Smart CCTV' Experiment," *Culture Unbound: Journal of Current Cultural Research* 2 (2010): 85.

14. See, for instance, Timothy Lenz, *Changing Images of Law in Film and Television Crime Stories* (New York: Peter Lang, 2003); and the essays in the collected volume *Entertaining Crime: Television Reality Programs,* ed. Mark Fishman and Gray Cavendar (Piscataway, N.J.: Aldine Transactions, 1998).

15. Battles, *Calling All Cars.*

16. Mariana Valverde, *Law and Order: Images, Meanings, Myths* (New Brunswick, N.J.: Rutgers University Press, 2006).

17. Nicole Hahn Rafter, *Shots in the Mirror: Crime Films and Society* (Oxford, UK: Oxford University Press, 2006).

18. See, for example, Mark Andrejevic, *Reality TV: The Work of Being Watched* (Lanham, Md.: Rowman and Littlefield, 2003); Ronald Walter Greene, "Lessons from the YMCA: The Material Rhetoric of Criticism, Rhetorical Interpretation, and Pastoral Power," in *Communication Matters: Materialist Approaches to Media, Mobility, and Networks,* ed. Jeremy Packer and Stephen B. Crofts Wiley (London: Routledge, 2012), 219–30; Laurie Oullette and James Hay, *Better Living through Reality TV: Television and Post-welfare Citizenship* (Malden, Mass.: Blackwell, 2008); and Shayne Pepper, "Public Service Entertainment: Post-network Television, HBO, and the AIDS Epidemic" (PhD diss., North Carolina State University, 2011).

19. James Hay has provided a macro-scale approach to media as technologies of liberal government in his investigation of the historical role of cinema, radio, and television as mechanisms for "the spatial rationalization of bodies, movements, knowledge, and observation" in succeeding urban renewal initiatives in the United States. Such an approach is more in line with ours, though Hay's media of choice are mass-media whereas we focus on media explicitly designed or applied to policing in its narrow sense. See Hay, "The Birth of the 'Neoliberal' City and Its Media," in *Communication Matters: Materialist Approaches to Media, Mobility, and Networks,* ed. Jeremy Packer and Steven B. Crofts Wiley (London: Routledge, 2012), 126.

20. Michel Foucault, "Space, Knowledge, Power," in *The Foucault Reader,* ed. Paul Rabinow (New York: Pantheon 1984), 241.

21. Foucault, 242.

22. See Joshua Reeves, *Citizen Spies: The Long Rise of America's Surveillance Society* (New York: New York University Press, 2017); and Foucault, "Space, Knowledge, Power," 242.

23. Michel Foucault, *Security, Territory, Population: Lectures at the Collège de France, 1977–1978,* ed. Michael Senellart, trans. Graham Burchell (London: Palgrave, 2007), 27.

24. Foucault, "Space, Knowledge, Power," 244.

25. See Foucault, *Security, Territory, Population.* As we address later in the chapter, this trajectory has often been described as the shift from disciplinarity to governmentality. Regardless of the specific vocabulary, Foucault explained that there have not been simple replacements of one form of power by another (e.g., sovereignty replaced by disciplinarity replaced by governmentality), but rather that these three become fluid, with their interplay leading to one becoming more

or less predominant at a given time. See *Security, Territory, Population*, 106–8. For a more thorough discussion of the relationship between discipline, security, and control, see Greg Elmer, "Panopticon—Discipline—Control," in *The Routledge Handbook of Surveillance Studies*, ed. Kirstie Ball, Kevin Haggerty, and David Lyon (New York: Routledge, 2012), 21–29.

26. Foucault, *Security, Territory, Population*, 44.

27. Foucault, 45.

28. Foucault, 45.

29. Foucault, 48.

30. Foucault, 49.

31. Foucault, 34, 71.

32. Stan Allen, "Infrastructural Urbanism," in *Scroope 9* (Cambridge, UK: Cambridge University Architecture School, 1998), 71–79.

33. Michel Foucault, *"Society Must Be Defended": Lectures at the Collège de France, 1975–1976* (New York: Palgrave, 2003), 134.

34. Foucault, 134.

35. Foucault, 102.

36. Michel Foucault, *Discipline and Punish: The Birth of the Prison*, trans. Alan Sheridan (New York: Pantheon, 1977), 90–93, 104–7. Also see especially Foucault's "Truth and Juridical Forms" in *Power: Essential Works of Foucault*, vol. 3, ed. James D. Faubion (New York: Norton, 2000), 1–88.

37. Cesare Beccaria, *On Crimes and Punishments and Other Writings*, trans. Aaron Thomas and Jeremy Parzen, ed. Aaron Thomas (Toronto: University of Toronto Press, 2008), 79.

38. Beccaria, 79–80.

39. Beccaria, 79.

40. H. L. A. Hart, *Essays on Bentham: Studies in Jurisprudence and Political Theory* (Oxford: Clarendon Press, 1982).

41. For a fuller description of this medieval watch-and-ward policing method, see Joshua Reeves, "Lateral Surveillance and the Uses of Responsibility," *Surveillance & Society* 10, nos. 3–4 (2012): 240–42.

42. Critchley, *A History of Police*, 42.

43. See Colquhoun's *Treatise on the Police of the Metropolis* (London, 1796), where he bemoans the "absurd prejudice" that pitted a reluctant general public against the fledgling police force: "that the best laws that ever were made can avail nothing, if the public mind is impressed with an idea that it is a matter of infamy to become the casual or professional agent to carry them into execution" (213–14).

44. For more on how communication technologies are used to organize police patrols across time and space, see Richard V. Ericson and Kevin D. Haggerty, *Policing the Risk Society* (New York: Oxford University Press, 1997), 388–411; and Peter K. Manning, *The Technology of Policing: Crime Mapping, Information*

Technology, and the Rationality of Crime Control (New York: New York University Press, 2013).

45. Patrick Colquhoun, *Treatise on the Commerce and Police of the River Thames, Containing a Historical View of the Trade of the Port of London* (London, 1800), 41.

46. Colquhoun, 281.

47. David Garland, *Culture of Control: Crime and Social Order in Contemporary Society* (Chicago: University of Chicago Press, 2002), 31.

48. L. J. Hume, *Bentham and Bureaucracy* (Cambridge: Cambridge University Press, 1981), 114.

49. For a theoretical overview of lateral surveillance and a description of its contemporary trends, see Mark Andrejevic, "The Work of Watching One Another: Lateral Surveillance, Risk, and Governance," *Surveillance and Society* 2 (2005): 479–97.

50. Colquhoun, *Treatise on the Police of the Metropolis,* 101.

51. Critchley, *A History of Police,* 61.

52. Colquhoun, *Treatise on the Commerce and Police of the River Thames,* 101, 104.

53. See Hume, *Bentham and Bureaucracy,* 152.

54. Jeremy Bentham, *The Works of Jeremy Bentham*, vol. 8, ed. John Bowring (London: Simpkin, Marshall, and Co., 1843), 392, quoted in Hume, *Bentham and Bureaucracy*, 152.

55. *First Report of the Commissioners Appointed to Inquire as to the Best Means of Establishing an Efficient Constabulary Force in the Counties of England and Wales* (London: W. Clowes and Sons, 1839), 69.

56. The hue-and-cry policing method was the foundation of medieval crime response in England. Once a crime was committed, witnesses would raise a "hue and cry"—utilizing their voices, whistles, and whatever else they had on hand—and would trail the criminal until s/he was captured. Thus every male citizen could be deputized into an ad hoc police force at any moment. For an overlook of hue and cry and other premodern policing practices, see Critchley, *A History of Police,* 1–28, and Lucia Zedner, "Policing before and after the Police: The Historical Antecedents of Contemporary Crime Control," *British Journal of Criminology* 46 (2006): 78–96.

57. See Leon Radzinowicz, *A History of English Criminal Law and Its Administration from 1770* (New York: Macmillan, 1957), 47–50.

58. J. T. Barber Beaumont, "Regulations of Police for the Prevention of Disorders and Violence," in *The Pamphleteer: Dedicated to Both Houses of Parliament,* vol. 19 (London, 1822), 131.

59. Nathaniel Conant, "Minutes of Evidence Taken before the Committee on

the State of the Police of the Metropolis," in *Report from the Committee on the State of the Police of the Metropolis* (London, 1816), 16.

60. "Much of the lengthy intellectual history of criminology," writes Piers Beirne, "has been dominated by the belief that physical features are external signs of inner and spiritual darkness." Beirne, *Inventing Criminology: Essays on the Rise of "Homo Criminalis"* (Albany: SUNY Press, 1993), 187. Beirne's work traces the historical development of *Homo criminalis* in the criminological imagination, showing how typologies of the criminal were especially dominant in the nineteenth century and accompanied sociological attempts—such as those by Adolphe Quetelet—to theorize the "common man" (1–6).

61. For work on the relationship between media—particularly photography—and eugenics, see Allan Sekula, "The Body and the Archive," in *The Contest of Meaning: Critical Histories of Photography,* ed. Richard Bolton (Cambridge: MIT Press, 1989), 343–87; and Phillip Prodger, *Darwin's Camera: Art and Photography in the Theory of Evolution* (New York: Oxford University Press, 2009).

62. Robert Wilson McClaughry, "Publisher's Preface," in Alphonse Bertillon, *Signaletic Instructions: Including the Theory and Practice of Anthropometrical Identification* (Chicago: Werner, 1896), vii.

63. See Kelly A. Gates, "Biometrics and Post-9/11 Technostalgia," *Social Text* 23, no. 2 (2005): 41. Also see Valverde, *Law and Order,* 74–76.

64. See Armand Mattelart, *The Globalization of Surveillance,* trans. Susan Taponier and James A. Cohen (Malden, Mass.: Polity, 2010), 15–17.

65. Bertillon, *Signaletic Instructions,* 249–58.

66. Thomas Byrnes, *Professional Criminals of America* (New York: Cassell and Company, 1886).

67. Benjamin P. Eldridge, *Our Rival the Rascal: A Faithful Portrayal of the Conflict between the Criminals of This Age and the Defenders of Society—the Police* (Boston: Pemberton, 1896), 321.

68. Byrnes, *Professional Criminals of America,* 55.

69. Byrnes, preface.

70. As Critchley points out, these innovations retained their prominence in twentieth-century policing: "From the early beginnings the forensic laboratory system has grown until today it is an integral part of the police service, employing pathologists, chemists, biologists, experts in hand-writing, and many others. Nowhere else, perhaps, is the quiet drama of police work so vividly presented to the layman as in these quiet laboratories, where the examination of blood-stained sheets, the comparison of hairs and bits of skin, the analysis of human organs pickled in jars, and the microscopic examination of stains, specimens, and minute tell-tale traces of all kinds from the scenes of innumerable crimes make up the daily work" (*A History of Police,* 213–14).

71. See Jeremy Packer and Joshua Reeves, "Making Enemies with Media," *Communication and the Public* 5, nos. 1–2 (2020): 16–25.

72. See Gary T. Marx, *Undercover: Police Surveillance in America* (Berkeley: University of California Press, 1988); also see Mark Andrejevic, *iSpy: Surveillance and Power in the Interactive Era* (Lawrence: University Press of Kansas, 2006), 124–25.

73. See, for instance, Charles Tempest Clarkson and J. Hall Richardson, *Police!* (London: Leadenhall Press, 1889), 280–85.

74. See Reeves, *Citizen Spies*, 25–33, for more information about media in hue-and-cry-based policing.

75. Critchley, *A History of Police*, 109–10. Telegraph-based fire alarm systems came into use around this time as well. See Stewart, "The Police Signal Box." See Reeves, *Citizen Spies*, 55–64.

76. G. Douglas Gourley and Allen P. Bristow, *Patrol Administration* (Springfield, Ill.: Charles C. Thomas, 1961), 179.

77. Gourley and Bristow, 179.

78. Gourley and Bristow stress that police supervision is necessary not only to maintain discipline but also to produce an extensive police record that can be used to determine future allocation of resources. Automating such surveillance through media is a common solution. Also see Ericson and Haggerty, *Policing the Risk Society*.

79. For decades, call boxes were accessible only to police. In the 1950s an increasing number of these boxes were made available to civilians as a direct means of communication to police headquarters. Call boxes remained necessary well after the advent of police radio, as they allowed for secrecy where radio failed. Further, they were less prone to the noise of bloated airways pushed beyond their bandwidth capacity. By the 1960s, call boxes were deemed economically inefficient due to the high density of public telephones (Gourley and Bristow, *Patrol Administration*). For more detail, see the second chapter of Reeves, *Citizen Spies*.

80. In different ways and for differing reasons, several scholars have suggested that automobiles might fruitfully be treated as media or communication technologies. See, for instance, Marshall McLuhan, *Understanding Media: The Extensions of Man* (New York: McGraw-Hill, 1964); James Hay and Jeremy Packer, "Crossing the Media(-n): Auto-mobility, the Transported Self, and Technologies of Freedom," in *MediaSpace: Place, Scale, and Culture in a Media Age*, ed. Nick Couldry and Anna McCarthy (New York: Routledge, 2004), 209–32; and Sarah Sharma, "Taxis as Media: A Tempero-Materialist Reading of the Taxi-Cab," *Social Identities* 14, no. 4 (2008): 457–64.

81. Roads, in fact, had been a special preoccupation of police since their inception. In 1839, ten years after Robert Peel spearheaded the modern police force in England, a commission convened to assess the effectiveness of the new police.

One of their major concerns was the insecurity of travelers on rural highways. See *First Report of the Commissioners Appointed to Inquire as to the Best Means of Establishing an Efficient Constabulary Force in the Counties of England and Wales*, 87–92.

82. With the advent of the patrol car, we see the police striving to maintain a speed/territory edge on criminals. This became particularly germane during the Prohibition Era, when bootleggers used increasingly faster cars for transporting alcohol. See Jeremy Packer, *Mobility without Mayhem: Cars, Safety, and Citizenship* (Durham, N.C.: Duke University Press, 2008). The patrol car has been widely studied and invested in: for instance, based upon an internal study of the percentage of successful automobile pursuits by six-cylinder versus eight-cylinder patrol cars, the city of Los Angeles in 1958 determined that they would use only eight-cylinder cars for patrol, while six-cylinder vehicles were delegated to investigation and transportation (Gourley and Bristow, *Patrol Administration*).

83. "The San Francisco Police Department: 150 Years of History," *S.F.Police.org*, http://sf-police.org/index.aspx?page=1592, last accessed May 18, 2012.

84. Paul Weston, *The Police Traffic Control Function* (Springfield, Ill.: Thomas, 1968), 3.

85. For a prewar analysis see David Blanke, *Hell on Wheels: The Promise and Peril of America's Car Culture, 1910–1940* (Lawrence: University Press of Kansas, 2007); and for a postwar analysis see Packer, *Mobility without Mayhem*.

86. See Jeremy Packer, "Rethinking Dependency: New Relations of Transportation and Communication," in *Thinking with James Carey: Essays on Communications, Transportation, History*, ed. Jeremy Packer and Craig Robertson (New York: Peter Lang 2006), 79–100.

87. See Weston, *The Police Traffic Control Function*. Also see Ross D. Petty, "The Rise, Fall, and Rebirth of Bicycle Policing," *International Police Mountain Bike Association*, 2006, http://www.ipmba.org/newsletters/ABriefHistoryof PoliceCycling.pdf.

88. See Rollo N. Harger, "Some Practical Aspects of Chemical Tests for Intoxication," *Journal of Criminal Law and Criminology* 35 (1944): 202–18.

89. R. F. Borkenstein and H. W. Smith, "The Breathalyzer and Its Applications," *Medicine, Science, and the Law* 2 (1961): 13–22.

90. Weston, *The Police Traffic Control Function*, 216–21.

91. Weston, 218.

92. Weston, 230.

93. Responses to the perceived urgency of traffic safety were prolific following the publication of Ralph Nader's *Unsafe at Any Speed: The Designed-in Danger of the American Automobile* (New York: Grossman, 1965) and led Congress to pass the Traffic and Motor Vehicle Safety Act.

94. Weston, *The Police Traffic Control Function*.

95. As Bruno Latour has argued, technologies like automatic door closers have agency and elicit specific forms of action from humans. See Latour, "Where Are the Missing Masses? The Sociology of a Few Mundane Artefacts," in *Shaping Technology/Building Society: Studies in Sociotechnical Change*, ed. Wiebe E. Bijker and John Law (Cambridge: MIT Press, 1988), 225–58.

96. The German blitzkrieg depended upon cryptographically encoded radio communications. See Friedrich Kittler, "Media Wars: Trenches, Lighting, Stars," in *Literature, Media, Information Systems: Essays* (Amsterdam: Overseas Publishers, 1997), 117–29. Aerial photographs were used to map and plan attacks beginning as early as the Civil War, but they gained prominence with film and airplanes during World War I. See Paul Virilio, *War and Cinema: The Logistics of Perception* (New York: Verso, 1989).

97. United States Government Accountability Office, "Testimony before the Committee on Government Reform, House of Representatives," in *9/11 Commission Report: Reorganization, Transformation, and Information Sharing*, 2004, http://www.gao.gov/new.items/d041033t.pdf.

98. "Department of Justice Information Technology Strategic Plan, 2008–2013," *Justice.gov* (2008): 16, https://www.justice.gov/archive/jmd/irm/2008itplan/p3.pdf.

99. United States Department of Justice, "Department of Justice IT Strategic Plan Fiscal Years 2010–2015," *Justive.gov*, December 9, 2009, http://www.justice.gov/jmd/ocio/it-strategic-plan.htm.

100. See Alexander Galloway and Eugene Thacker, *The Exploit: A Theory of Networks* (Minneapolis: University of Minnesota Press, 2007).

101. National Institute of Justice, "Effective Police Communications Systems Require New 'Governance,'" June 2008, *NIJ.org*, http://www.nij.gov/topics/technology/communication/governance.htm.

102. Torin Monahan, "The Future of Security? Surveillance Operations at Homeland Security Fusion Centers," *Social Justice* 37, nos. 2–3 (2010): 84–98.

103. See Greg Elmer, "Policing Space and Race: Prejudice, Policies, and Racial Profiling," *Space and Culture* 11/12 (2001): 179.

104. Elmer, "Policing Space and Race"; and David Lyon, *Surveillance Studies: An Overview* (Malden, Mass.: Polity, 2007), 201.

105. Gary T. Marx, "Soft Surveillance: The Growth of Mandatory Volunteerism in Collecting Personal Information—'Hey Buddy Can You Spare a DNA?,'" in *Surveillance and Security: Technological Politics and Power in Everyday Life*, ed. Torin Monahan (New York: Routledge, 2006), 37–56.

106. Friedrich Kittler, *Optical Media: Berlin Lectures 1999*, trans. Anthony Enns (Malden, Mass.: Polity, 2010).

107. For an analysis of how such policing came into being, see Ericson and Haggerty, *Policing the Risk Society*.

108. Jack Bratich, "User-Generated Discontent: Convergence, Polemology, and Dissent," *Cultural Studies* 25, nos. 4–5 (2011): 621–40.

109. See Torin Monahan, "Counter-Surveillance as Political Intervention," *Social Semiotics* 16, no. 4 (2006): 515–34. Also see Glencora Borradaile and Joshua Reeves, "Sousveillance Capitalism," *Surveillance & Society* 18, no. 2 (2020): 272–75.

110. Greg Elmer and Andy Opel, *Preempting Dissent: The Politics of an Inevitable Future* (Winnipeg, Manitoba: Arbeiter Ring, 2008).

111. Patrick F. Gillham, Bob Edwards, and John A. Noakes, "Strategic Incapacitation and the Policing of Occupy Wall Street Protests in New York City, 2011," *Policing and Society* 23 (2013): 81–102. Also see Bratich, "User-Generated Discontent."

112. "Chicago Police Sound Cannon: LRAD 'Sonic Weapon' Purchased Ahead of NATO Protests," *Huffington Post*, May 15, 2012, http://www.huffingtonpost.com/2012/05/15/chicago-police-sound-cannon-lrad-nato-summit_n_1518322.html; also see Daphne Carr, "Understanding the LRAD, the 'Sound Cannon' Police Are Using at Protests, and How to Protect Yourself from It," *Pitchfork Media*, June 9, 2020, https://pitchfork.com/thepitch/understanding-the-lrad-the-sound-cannon-police-are-using-at-protests-and-how-to-protect-yourself-from-it/.

4. Circuitous Maximus

1. A phrase from the days of the railroad, a steam-powered circuit.

2. John Humphrey, *Roman Circuses: Arenas for Chariot Racing* (Berkeley: University of California Press, 1986), 73.

3. Raymond Williams, *Television: Technology and Cultural Form*, ed. Ederyn Williams (New York: Routledge, 2003).

4. Jeremy Packer and Kathleen Oswald, "From Windscreen to Widescreen: Screening Technologies and Mobile Communication," *Communication Review* 13, no. 4 (2010): 309–39; Kathleen Oswald and Jeremy Packer, "Flow and Mobile Media: Broadcast Fixity to Digital Fluidity," in *Communication Matters: Materialist Approaches to Media, Mobility, and Networks*, ed. Jeremy Packer and Stephen B. Crofts Wiley (Abingdon, UK: Routledge, 2012), 276–87.

5. Colin Koopman, in "Problematization in Foucault's Genealogy and Deleuze's Symptomatology: Or, How to Study Sexuality without Invoking Oppositions," *Angelaki* 23, no. 2 (2018): 187–204, suggests that problematization in the work of Foucault and Deleuze is distinguished by a mode of philosophy that is "critical (rather than metaphysical), immanent (rather than transcendental), experimental (rather than dialectical), and problematizing (rather than problem-solving) mode of philosophy" (188). While Williams is not aiming to carry out a genealogical analysis, his approach in *Television* resonates with such an approach.

6. He calls out McLuhan specifically.

7. Williams, *Television,* 13.

8. Williams, 13.

9. Friedrich Kittler, *Optical Media* (Malden, Mass.: Polity, 2010), following Claude Shannon, "A Mathematical Theory of Communication," *Bell System Technical Journal* 27 (1948): 379–423. As Shannon suggested, "The significant aspect is that the actual message is one selected from a set of possible messages" (379).

10. See William Boddy, "Interactive Television and Advertising Form in Contemporary US Television," in *Television after TV: Essays on a Medium in Transition,* ed. Lynn Spigel and Jan Olsson (Durham, N.C.: Duke University Press, 2004), 113–32; James Hay, "The (Neo)Liberalization of the Domestic Sphere and the New Architecture of Community," in *Foucault, Cultural Studies, and Governmentality*, ed. Jack Bratich, Jeremy Packer, and Cameron McCarthy (Albany: SUNY Press, 2003), 165–206; Michael Kackman, Marine Binfield, Matthew Thomas Payne, Allison Perlman, and Bryan Sebok, eds., *Flow TV: Television in the Age of Media Convergence* (New York: Routledge, 2010); Lynn Spigel and Jan Olsson, eds., *Television after TV: Essays on a Medium in Transition* (Durham, N.C.: Duke University Press, 2004), 2.

11. Williams, *Television,* 16.

12. Lynn Spigel, *Make Room for TV: Television and the Family Ideal in Postwar America* (Chicago: University of Chicago Press, 1992).

13. Mark Andrejevic, in *Reality TV: The Work of Being Watched* (Lanham, Md.: Rowman and Littlefield, 2003); and *iSpy: Surveillance and Power in the Interactive Era* (Lawrence: University of Kansas Press, 2009), suggests that watching and being watched are both work done to generate capital for media conglomerates. From this perspective, flow provides an explanation of how time could be made most productive for the longest period of time.

14. Sut Jhally and Bill Livant, "Watching as Working: The Valorization of Audience Consciousness," *Journal of Communication* 36, no. 3 (1986): 124–43, inspired by Dallas W. Smythe, "On the Audience Commodity and Its Work," in *Dependency Road: Communications, Capitalism, Consciousness, and Canada* (Norwood, N.J.: Ablex, 1981), 22–51.

15. Jeremy Packer, "Mobile Communications and Governing the Mobile: CBs and Truckers," *Communication Review* 5, no. 1 (2002): 39–57.

16. Jonathan Sterne, "Sounds Like the Mall of America: Programmed Music and the Architectonics of Commercial Space," *Ethnomusicology* 41, no. 1 (1997): 22–50.

17. Raymond Williams and Michael Orrom, *Preface to Film* (London: Film Drama, 1954).

18. Raymond Williams, *Culture and Society, 1780–1950* (London: Chatto & Windus, 1958).

19. Vincent Mosco, "After the Internet: New Technologies, Social Issues, and Public Policies," *Fudan Journal of the Humanities and Social Sciences* 10, no. 3 (2017): 297–33; Mark Andrejevic and Mark Burdon, "Defining the Sensor Society," *Television and New Media* 16, no. 1 (2015): 19–36; Orit Halpern, *Beautiful Data: A History of Vision and Reason since 1945* (Durham, N.C.: Duke University Press, 2015); Danah Boyd and Kate Crawford, "Critical Questions for Gig Data," *Information, Communication & Society* 15, no. 5 (2012): 662–79; Shannon Mattern, "Databodies in Codespace," *Places*, 2018, https://placesjournal.org/article/databodies-in-codespace/.

20. Mikey Campbell, "Trump Refuses to Give Up iPhones, Chinese and Russian Spies Eavesdrop on Calls," *AppleInsider*, October 25, 2018.

21. See https://www.nielsen.com/content/dam/corporate/us/en/reports-downloads/2019-reports/q3-2018-total-audience-report.pdf. The numbers have hovered around eight hours for the past several years, but they change as different devices/contexts are entered into the calculations.

22. Michel Foucault, *Discipline and Punish: The Birth of the Prison*, trans. Alan Sheridan (New York: Pantheon, 1977).

23. We fully acknowledge that these are not fixed categories, but recognize that they are often treated as rigid categories by sorting mechanisms such as surveys, school records, and border agents.

24. Andrejevic and Burdon, "Defining the Sensor Society," 21.

25. Packer and Oswald, "From Windscreen to Widescreen."

26. Mosco, "After the Internet."

27. Friedrich Kittler, *Gramophone, Film, Typewriter* (Stanford: Stanford University Press, 1999).

28. This description is meant to emphasize the increasingly controlled experience of the driver. We do not mean, of course, that the digital is in fact immaterial. The genealogical perspective reveals quite the opposite.

29. Paulo Silva and Johan Huijsing, *High-Resolution IF-to-Baseband SigmaDelta ADC for Car Radios* (Berlin: Springer Science & Business Media, 2008), 5.

30. While an audiophile LP can weigh upwards of 180 grams, be stored in a non-gatefold sleeve that measures 12.25 by .125 inches, and averages 45 minutes of audio content, the (2010) version of Apple's iPod Shuffle weighed less than 11 grams, is 1.8 x 0.7 x 0.3 inches and can hold approximately 4,000 minutes of audio content. The weight-to-data ratio for the Shuffle is about 1,600 times more efficient than the high-quality LP as a storage device. Today's always accessible music streaming platforms bring libraries to users that blow the iPod out of the water.

31. For the user, of course. These systems are totally material and suck down juice like a black hole from sensor to server and back again.

32. Such as the billboard; see Catherine Gudis, *Buyways: Billboards, Automobiles, and the American Landscape* (New York: Routledge, 2004).

33. See Christopher Finch, *Highways to Heaven: The Auto Biography of America* (New York: Perennial, 1993).

34. Robert Venturi, Steven Izenour, and Denise Scott Brown, *Learning from Las Vegas: The Forgotten Symbolism of Architectural Form* (Cambridge: MIT Press, 1977).

35. Lynn Spigel, "Media Homes: Then and Now," *International Journal of Cultural Studies* 4, no. 4 (2001): 391.

36. Spigel, 392.

37. See Peter Norton's *Fighting Traffic: The Dawn of the Motor Age in the American City* (Cambridge: MIT Press, 2008).

38. Cell phones are known as such due to the organization of the physical infrastructure resulting in "cells" rather than coming from something that has actually to do with the device.

39. See Alessandra Renzi and Greg Elmer, "The Biopolitics of Sacrifice: Securing Infrastructure at the G20 Summit," *Theory, Culture, & Society* 30, no. 5 (2013): 49–51.

40. Copper Development Association Inc., https://www.copper.org/consumers/copperhome/Technology/.

41. Justin Stoltzfus. "Your Car, Your Computer: ECUs and the Controller Area Network," *Technopedia*, August 7, 2020, https://www.techopedia.com/your-car-your-computer-ecus-and-the-controller-area-network/2/32218.

42. Martin Placek, "Automotive Electronics Cost as a Share of Total Car Cost from 1970 to 2030," *Statista*, February 4, 2022, https://www.statista.com/statistics/277931/automotive-electronics-cost-as-a-share-of-total-car-cost-worldwide/.

43. Carolyn Mathas, "The Price Tag of Automotive Electronics: What's Really at Play?," *EDN: Electronics Design Network*, September 25, 2017, https://www.edn.com/the-price-tag-of-automotive-electronics-whats-really-at-play/.

44. Austin Weber, "Wire's Role in a Drive-by-Wire World," *Assembly*, May 27, 2010, https://www.assemblymag.com/articles/87479-wire-s-role-in-a-drive-by-wire-world.

45. Weber, para. 4.

46. Simon Dixon, Haris Irshad, Derek M. Pankratz, and Justine Bornstein, "The 2019 Deloitte City Mobility Index," *Deloitte Insights*, https://www2.deloitte.com/content/dam/Deloitte/br/Documents/consumer-business/City-Mobility-Index-2019.pdf.

47. Jon Markman, "Microsoft Drives Stunning Success in New World of Connected-Car Platforms," *Forbes*, March 23, 2019, https://www.forbes.com/sites/jonmarkman/2019/03/23/microsoft-drives-stunning-success-in-new-world-of-connected-car-platforms/#b90b70472450.

48. KPMG, "Reimagine Places: Mobility as a Service," August 2017, https://assets.kpmg/content/dam/kpmg/uk/pdf/2017/08/reimagine_places_maas.pdf.

49. Frost & Sullivan, "Global Autonomous Driving Market Outlook, 2018," March 2018, Global Automotive & Transportation Research Team at Frost & Sullivan, slide 19, http://info.microsoft.com/rs/157-GQE-382/images/K24A-2018%20Frost%20%26%20Sullivan%20-%20Global%20Autonomous%20Driving%20Outlook.pdf.

50. Edmunds, "2019 Forecast and Trends," December 18, 2018, https://static.ed.edmunds-media.com/unversioned/img/industry-center/analysis/2019-edmunds-forecast-and-trends.pdf.

51. https://static.ed.edmunds-media.com/unversioned/img/industry-center/analysis/2019-edmunds-forecast-and-trends.pdf.

52. Fortune Business Insights, "Transportation & Logistics, Car Leasing Market," https://www.fortunebusinessinsights.com/car-leasing-market-105417.

53. Toni Fitzgerald, "How Many Streaming Video Services Does the Average Person Subscribe To?," *Forbes*, March 29, 2019, https://www.forbes.com/sites/tonifitzgerald/2019/03/29/how-many-streaming-video-services-does-the-average-person-subscribe-to/.

54. James Anthony, "74 Amazon Statistics You Must Know: 2021/2022 Market Share Analysis & Data," *Finances Online,* https://financesonline.com/amazon-statistics/.

55. Edmunds, "Transportation Transformation: The Current State of the Autonomous Car," October 2017, https://static.ed.edmunds-media.com/unversioned/img/industry-center/analysis/autonomous-report.pdf.

56. McKinsey & Company, "McKinsey Electric Vehicle Index: Europe Cushions a Global Plunge in EV Sales," July 17, 2020, https://www.mckinsey.com/industries/automotive-and-assembly/our-insights/mckinsey-electric-vehicle-index-europe-cushions-a-global-plunge-in-ev-sales.

57. IEA, "Global EV Outlook 2021," April 2021, https://www.iea.org/reports/global-ev-outlook-2021/trends-and-developments-in-electric-vehicle-markets#abstract.

58. "Transportation Transformation: The Current State of the Autonomous Car," *Edmunds*, October 2017.

59. Nicholas Chase, John Males, and Mark Schipper, "Autonomous Vehicles: Uncertainties and Energy Implications," *EIA Annual Energy Outlook 2018,* May 30, 2018, 4, https://www.eia.gov/outlooks/aeo/av.php.

5. How We All Were Committed

1. Hannah Muniz Castro et al., "A Case of Attempted Bilateral Self-Enucleation in a Patient with Bipolar Disorder," *Mental Illness* 9, no. 1 (2017): 7141.

2. Nicholas P. Jones, "Self-Enucleation and Psychosis," *British Journal of Ophthalmology* 74 (1990): 571–73.

3. Jones, 572.

4. Jones, 572.

5. Jones, 572.

6. "Vision seems to play a prominent role in the development of psychosis given that basic visual symptoms identified before illness onset are one of the most powerful predictors of the emergence of later psychotic disorders." Christoph Teufel, Naresh Subramaniam, Veronika Dobler, Jesus Perez, Johanna Finnemann, Puja R. Mehta, Ian M. Goodyer, and Paul C. Fletcher, "Shift toward Prior Knowledge Confers a Perceptual Advantage in Early Psychosis and Psychosis-Prone Healthy Individuals," *Proceedings of the National Academy of Sciences of the United States of America* 112, no. 43 (2015): 13401, https://doi.org/10.1073/pnas.1503916112.

7. Joachim Klosterkötter, Martin Hellmich, E. M. Steinmeyer, and Frauke Schultze-Lutter, "Diagnosing Schizophrenia in the Initial Prodromal Phase," *Archives of General Psychiatry* 58, no. 2 (2001): 158–64.

8. Esmé Weijun Wang, author of *The Collected Schizophrenias*, provided this poignant summation of hallucinatory experience during a radio interview for National Public Radio. See https://www.npr.org/2019/02/03/690501557/hallucinations-kidnap-the-senses-in-the-collected-schizophrenias.

9. Flavie Waters and Charles Fernyhough, "Hallucinations: A Systematic Review of Points of Similarity and Difference across Diagnostic Classes," *Schizophrenia Bulletin* 43, no. 1 (2017): 32–43.

10. Plato, *The Republic of Plato* (London: Oxford University Press, 1945).

11. Teufel et al., "Shift toward Prior Knowledge," 13401.

12. Teufel et al., 13402.

13. Teufel et al., 13405.

14. See Friedrich A. Kittler, *Optical Media: Berlin Lectures 1999*, trans. Anthony Enns (Malden, Mass.: Polity, 2010).

15. "Spectacle," The University of Chicago Glossary, Theories of Media: Keyword Glossary, http://csmt.uchicago.edu/glossary2004/spectacle.htm.

16. Paolo Legrenzi, "Spectacles: Experience, Science, and Imagination," in *Taking Glasses Seriously: Art, History, Science, and Technologies of the Vision*, ed. Raimonda Riccini (Milan: La Triennale di Milano, 2002), 48.

17. Marshall McLuhan, *Understanding Media: The Extensions of Man*, ed. Scott Boms (Berkeley, Calif.: Gingko Press, 2013), Kindle location 9.

18. Orit Halpern, "Perceptual Machines: Communication, Archiving, and Vision in Post-war American Design," *Journal of Visual Culture* 11, no. 3 (December 1, 2012): 328–51, https://doi.org/10.1177/1470412912455619.

19. Claude E. Shannon and Warren Weaver, *The Mathematical Theory of Communication* (Urbana: University of Illinois Press, 1975).

20. Shannon and Weaver.

21. Kittler, *Optical Media*, 34.

22. Kittler, 53.

23. Kittler, 36.

24. Kittler, 56.

25. "The people who standout as living in a nonperspective world are the Zulus (11) living in a world of 'circular culture,' with round huts, plowing land in curves, with few straight lines or corners at all." Marilita M. Moschos, "Physiology and Psychology of Vision and Its Disorders: A Review." *Medical Hypothesis Discovery and Innovation in Ophthalmology* 3, no. 3 (2014): 83–90.

26. Kittler, *Optical Media*, 57.

27. Marco Piccolino and Andrea Moriondo, "Retina and Vision: The Imperfect Image," in Riccini, *Taking Eyeglasses Seriously*, 133.

28. Orit Halpern, "Dreams for Our Perceptual Present: Temporality, Storage, and Interactivity in Cybernetics," *Configurations* 13, no. 2 (2005): 283–319, https://doi.org/10.1353/con.2007.0016.

29. Piccolino and Moriondo, "Retina and Vision," 133.

30. Piccolino and Moriondo, 133.

31. Shannon and Weaver, *The Mathematical Theory of Communication*.

32. Paul Rea, "Optic Nerve," in *Clinical Anatomy of the Cranial Nerves* (London: Elsevier, 2014), 7–26, https://doi.org/10.1016/B978-0-12-800898-0.00002-6.

33. Shannon and Weaver, *The Mathematical Theory of Communication*.

34. Matthew Schmolesky, "The Primary Visual Cortex," in *Webvision: The Organization of the Retina and Visual System*, ed. Helga Kolb, Eduardo Fernandez, and Ralph Nelson (Salt Lake City: University of Utah Health Sciences Center, 1995), http://www.ncbi.nlm.nih.gov/books/NBK11524/.

35. Simon Murray and Nicky Albrechtsen, *Fashion Spectacles, Spectacular Fashion: Eyewear Styles and Shapes from Vintage to 2020* (London: Thames & Hudson, 2012).

36. Melvin L. Rubin, "Spectacles: Past, Present, and Future," *Survey of Ophthalmology* 30, no. 5 (March 1986): 321–27, https://doi.org/10.1016/0039-6257(86)90064-0.

37. Murray and Albrechtsen, *Fashion Spectacles;* Rubin, "Spectacles."

38. Murray and Albrechtsen, *Fashion Spectacles;* Rubin, "Spectacles."

39. Murray and Albrechtsen, *Fashion Spectacles;* Rubin, "Spectacles."

40. Murray and Albrechtsen, *Fashion Spectacles;* Rubin, "Spectacles."

41. Bacon quoted in Murray and Albrechtsen, *Fashion Spectacles;* Rubin, "Spectacles."

42. Kittler, *Optical Media,* 36.

43. It should be noted that while the corrective lens was an object of the hand and not of the face, it was afforded less power in terms of managing the self due to the distance between lens and eye, which allowed for a greater field of view and therefore more unfiltered information.

44. Kittler, *Optical Media,* 36.

45. Tomás Maldonado, "Taking Eyeglasses Seriously," *Design Issues* 17, no. 4 (2001): 38.

46. Vincent Ilardi, *Renaissance Vision from Spectacles to Telescopes* (Philadelphia: American Philosophical Society, 2007), 25.

47. Ilardi.

48. Ilardi.

49. Ilardi, 4.

50. Ilardi, 8.

51. Ilardi.

52. Ilardi.

53. Ilardi.

54. Ilardi.

55. Ilardi, 86.

56. Ilardi, 53.

57. Ilardi, 30.

58. Ilardi.

59. Raimonda Riccini, "From the Naked Eye to the Electronic Eye," in Riccini, *Taking Eyeglasses Seriously,* 16.

60. Maldonado, "Taking Eyeglasses Seriously."

61. Umberto Eco, *The Name of the Rose,* Kindle edition (Boston: Houghton Mifflin Harcourt, n.d.).

62. Eco, Kindle locations 1161–63.

63. Robert Darnton, "History of Reading," in *New Perspective on Historical Writing,* ed. Peter Burke, 2nd ed. (University Park: Pennsylvania State University Press, 2001), 176.

64. Ilardi, *Renaissance Vision.*

65. Ilardi, 22.

66. Eltjo Buringh and Jan Luiten Van Zanden, "Charting the 'Rise of the West': Manuscripts and Printed Books in Europe, A Long-Term Perspective from the Sixth through Eighteenth Centuries," *Journal of Economic History; Santa Clara* 69, no. 2 (June 2009): 409–45, http://dx.doi.org.myaccess.library.utoronto.ca/10.1017/S0022050709000837.

67. Buringh and Van Zanden.

68. Eco, *The Name of the Rose,* Kindle locations 1158–60.

69. Ilardi, *Renaissance Vision.*

70. Ilardi. Also Murray and Albrechtsen, *Fashion Spectacles.*

71. Ilardi, *Renaissance Vision.*

72. Kittler, *Optical Media,* 66.

73. Murray and Albrechtsen, *Fashion Spectacles.*

74. Melvin L. Rubin, "Spectacles: Past, Present, and Future," *Survey of Ophthalmology* 30, no. 5 (March 1986): 321–27, https://doi.org/10.1016/0039-6257(86)90064-0.

75. Ilardi, *Renaissance Vision,* 77.

76. Legrenzi, "Spectacles."

77. Maldonado, "Taking Eyeglasses Seriously," 8.

78. Maldonado, 12.

79. Maldonado, 9.

80. Maldonado, 9.

81. Ilardi, *Renaissance Vision,* 79.

82. Ilardi, 88.

83. Ilardi.

84. Ilardi, 114.

85. Maldonado, "Taking Eyeglasses Seriously."

86. "The 'Inventor' of Bifocals?," June 13, 2011, https://web.archive.org/web/20110613044912/http://www.college-optometrists.org/en/knowledge-centre/museyeum/online_exhibitions/artgallery/bifocals.cfm.

87. "The 'Inventor' of Bifocals?"

88. "The 'Inventor' of Bifocals?"

89. Harold A. Stein, Raymond M. Stein, and Melvin I. Freeman, *The Ophthalmic Assistant: A Text for Allied and Associated Ophthalmic Personnel,* 9th ed. (Edinburgh: Saunders/Elsevier, 2013).

90. McLuhan, *Understanding Media,* Kindle location 366.

91. Amy L. Sheppard and James S. Wolffsohn, "Digital Eye Strain: Prevalence, Measurement and Amelioration," *BMJ Open Ophthalmology* 3, no. 1 (April 1, 2018): e000146, https://doi.org/10.1136/bmjophth-2018-000146.

92. Jeffrey I. Cole, Michael Suman, Phoebe Schramm, Liuning Zhou, Andromeda Salvador, Harlan Lebo, and Monica Dunahee, "Surveying the Digital Future" (University of Southern California, 2014), 163.

93. Riccini, "From the Naked Eye."

94. Sheppard and Wolffsohn, "Digital Eye Strain."

95. Sheppard and Wolffsohn.

96. L. Frank Baum, *The Master Key* (Project Gutenberg, 2008), Kindle Location 574.

97. https://patents.google.com/patent/US3050870?oq=3050870.

98. James E. Melzer and Kirk Wayne Moffitt, *Head-Mounted Displays: Designing for the User* (New York: McGraw-Hill, 1997), 2.

99. Melzer and Moffitt, 2.

100. Kittler, *Optical Media,* 36.

101. Isabel Pederson, *Ready to Wear: A Rhetoric of Wearable Computers and Reality-Shifiting Media* (Anderson, S.C.: Parlor Press, 2013), 39.

102. Kittler, *Optical Media,* 36.

103. Jorge Luis Borges, "Funes, the Memorious," in *The Argentina Reader,* ed. Gabriela Nouzeilles et al. (Durham, N.C.: Duke University Press, 2009), 309, https://doi.org/10.1215/9780822384182-050.

104. Borges, 310.

105. Borges, 311.

106. Orit Halpern, *Beautiful Data: A History of Vision and Reason since 1945* (Durham, N.C.: Duke University Press, 2014), 169.

107. Halpern, 201.

108. Halpern, 200.

109. Richard Lanham, *The Economics of Attention* (Chicago: University of Chicago Press, 2006), 6.

110. Mark Andrejevic, *Infoglut: How Too Much Information Is Changing the Way We Think and Know* (New York: Taylor & Francis, 2013), 2.

111. Mark Andrejevic, "Privacy, Exploitation, and the Digital Enclosure," *Amsterdam Law Forum* 1, no. 4 (2009): 53.

112. Nick Bilton, "Google Offers Look at Internet-Connected Glasses," *New York Times,* April 4, 2012, Technology section, https://www.nytimes.com/2012/04/05/technology/google-offers-look-at-internet-connected-glasses.html.

113. Mark Weiser, "The Computer for the 21st Century," *Scientific American* 265, no. 3 (1991): 94–104.

114. Weiser.

115. Friedrich A. Kittler, *Gramophone, Film, Typewriter,* trans. Geoffrey Winthrop-Young and Michael Wutz (Stanford: Stanford University Press, 1999), 3.

116. Edgar Allan Poe, "The Spectacles," in *The Edgar Allan Poe Collection* (London: Arcturus Publishing Limited, 2017).

117. "Keyboard—History of the Modern Computer Keyboard," https://history-computer.com/ModernComputer/Basis/keyboard.html.

118. "Keyboard."

119. Amit Pinchevski and John Durham Peters, "Autism and New Media: Disability between Technology and Society," *New Media & Society* 18, no. 11 (December 2016): 2507–23, https://doi.org/10.1177/1461444815594441.

120. "Visual Interpreting—Get Live, On-demand Access to Visual Information," https://aira.io/.

121. Pederson, *Ready to Wear,* 34.

122. Michel Foucault, *The History of Sexuality, Volume 3: The Care of the Self* (New York: Vintage Books, 1988), 62.

123. Kittler, *Gramophone, Film, Typewriter,* 3.
124. Kittler, 10.
125. Halpern, *Beautiful Data,* 210.
126. Halpern, 210.
127. Halpern, 211.
128. Wency Leung, "Record and Replay: How a Canadian-Made App Is Aiming to Help Alzheimer's Patients Improve Their Daily Lives," *Globe and Mail,* March 4, 2019, https://www.theglobeandmail.com/canada/article-toronto-teams-hippocamera-a-high-tech-memory-aid-for-alzheimers/.
129. Leung.
130. Thomas Way, Adam Bemiller, Raghavender Mysari, and Corinne Reimers, "Using Google Glass and Machine Learning to Assist People with Memory Deficiencies," *Int'l Conf. Artificial Intelligence,* 2015, 572, http://www.csc.villanova.edu/~tway/publications/Way_Bemiller_ICA6330.pdf.
131. Pederson, *Ready to Wear,* 100.
132. Way et al., "Using Google Glass," 574.
133. Way et al., 576.
134. Pederson, *Ready to Wear,* 106.
135. Pederson, 109.
136. Tracey Wallace, John T. Morris, and Scott Bradshaw, "EyeRemember: Memory Aid App for Google Glass," *Journal on Technology and Persons with Disabilities* 3 (2015): 118, 120.
137. Wallace, Morris, and Bradshaw, 120.
138. Foucault, *The History of Sexuality, Volume 3,* 50.
139. Pederson, *Ready to Wear.*
140. Michel Foucault, *Discipline and Punish: The Birth of the Prison,* trans. Alan Sheridan (New York: Pantheon, 1977), 200.
141. Michel Foucault, *The Order of Things: An Archaeology of the Human Sciences* (New York: Vintage Books, 1994), xix.
142. American Psychiatric Association, eds., *Diagnostic and Statistical Manual of Mental Disorders: DSM-5,* 5th ed. (Washington, D.C.: American Psychiatric Association, 2013), 31.
143. Pinchevski and Peters, "Autism and New Media," 2508.
144. Pinchevski and Peters, 2508.
145. "Glass Partners," Glass, https://x.company/glass/partners/.
146. "Brain Power," http://www.brain-power.com/.
147. Halpern, *Beautiful Data,* 168.
148. Catalin Voss et al., "Effect of Wearable Digital Intervention for Improving Socialization in Children with Autism Spectrum Disorder: A Randomized Clinical Trial," *JAMA Pediatrics* 173, no. 5 (2019): 446–54, https://doi.org/10.1001/jamapediatrics.2019.0285.

149. Voss et al.

150. Kittler, *Optical Media*, 228.

151. Oliver J. Muensterer, Martin Lacher, Christoph Zoeller, Matthew Bronstein, and Joachim Kübler, "Google Glass in Pediatric Surgery: An Exploratory Study," *International Journal of Surgery* 12, no. 4 (April 2014): 281–89.

152. "Chinese Police Unveil Camera Sunglasses," February 7, 2018, China section, https://www.bbc.com/news/world-asia-china-42973456.

153. "Chinese Police Are Using Smart Glasses to Identify Potential Suspects," *TechCrunch* (blog), accessed March 5, 2019, http://social.techcrunch.com/2018/02/08/chinese-police-are-getting-smart-glasses/.

154. "Chinese Police Unveil Camera Sunglasses."

155. "Chinese Police Unveil Camera Sunglasses."

156. "Azure IoT—Internet of Things Platform | Microsoft Azure," https://azure.microsoft.com/en-ca/overview/iot/.

157. Jessi Hempel, "Satya Nadella's Got a Plan to Make You Care about Microsoft. The First Step? Holograms," *Wired*, January 21, 2015, https://www.wired.com/2015/01/microsoft-nadella/.

158. Foucault, *The History of Sexuality, Volume 3*, 55.

159. Frederick Winslow Taylor, *The Principles of Scientific Management* (Mineola, N.Y.: Dover, 1997), 7.

160. Ganaele Langlois, *Meaning in the Age of Social Media* (New York: Palgrave Macmillan, 2014).

161. Hempel, "Satya Nadella's Got a Plan."

162. Pederson, *Ready to Wear*, 39.

163. "Glass Partners."

164. Larry Hardesty, "Extracting Audio from Visual Information: Algorithm Recovers Speech from the Vibrations of a Potato-Chip Bag Filmed through Soundproof Glass," *MIT News*, August 4, 2014, http://news.mit.edu/2014/algorithm-recovers-speech-from-vibrations-0804.

165. Michael Zhang, "The World's Fastest Camera Can Shoot 10 Trillion Frames Per Second," PetaPixel, October 15, 2018, https://petapixel.com/2018/10/15/the-worlds-fastest-camera-can-shoot-10-trillion-frames-per-second/.

6. Media Genealogical Method

1. Colin Koopman, *Genealogy as Critique: Foucault and the Problems of Modernity* (Bloomington: Indiana University Press, 2013).

2. Michel Foucault, *The Archaeology of Knowledge and the Discourse on Language*, trans. A. M. Sheridan Smith (New York: Pantheon, 1972), 166–67.

3. Michel Foucault, "The Archaeology of Knowledge," trans. John Johnston,

in *Foucault Live: Collected Interviews, 1961–1984,* ed. Sylvère Lotringer (New York: Semiotext(e), 1996), 59.

4. Foucault, *The Archaeology of Knowledge*, 234.

5. Hubert L. Dreyfus and Paul Rabinow, *Michel Foucault: Beyond Structuralism and Hermeneutics* (Chicago: University of Chicago Press, 1983).

6. Michael Behrent, "Genealogy of Genealogy: Foucault's 1970–1971 Course on the Will to Know," *Foucault Studies* 13 (2012): 157–78.

7. Behrent, 171.

8. The materiality of communication as well as the sensibility that power is best understood through an analysis of struggle will later come to the fore in the work of Friedrich Kittler.

9. Michel Foucault, "Nietzsche, Freud, Marx," in *Aesthetics, Method, and Epistemology: Essential Works of Foucault, 1954–1984,* ed. James D. Faubion and Paul Rabinow (New York: New Press, 1999), 269–78.

10. Michel Foucault, "Nietzsche, Genealogy, History," in *Aesthetics, Method, and Epistemology,* 369–92.

11. Foucault, 378.

12. Foucault, 377.

13. Foucault, 378.

14. See Foucault, 378–79.

15. Michel Foucault, "Return to History," in Foucault, *Aesthetics, Method, and Epistemology*, 430.

16. Michel Foucault, *Discipline and Punish: The Birth of the Prison,* trans. Alan Sheridan (New York: Vintage Books, 1995).

17. Foucault.

18. Dreyfus and Rabinow, *Michel Foucault,* 119.

19. Michel Foucault, "Clarifications on the Question of Power," trans. James Cascaito, in Lotringer, *Foucault Live,* 262.

20. Koopman, *Genealogy as Critique.*

21. Michel Foucault, "Problematics," in Lotringer, *Foucault Live,* 421.

22. Koopman, *Genealogy as Critique,* 48.

23. Koopman, 107.

24. Kittler's theoretical project will be described in detail later in this chapter.

25. Foucault, *Discipline and Punish,* 166.

26. John Durham Peters, *Speaking into the Air* (Chicago: University of Chicago Press, 2001).

27. John Nerone, "Approaches to Media History," in *A Companion to Media Studies,* ed. Angharad N. Valdivia (Malden, Mass.: Blackwell, 2003), 93–114.

28. Simon During, Introduction to *The Cultural Studies Reader,* ed. Simon During (New York: Routledge, 2007), 1–32.

29. Stuart Hall and Tony Jefferson, eds., *Resistance through Rituals: Youth Subcultures in Post-war Britain* (New York: Routledge, 2006).

30. See, e.g., John Hagan, *Learning to Labor: How Working Class Kids Get Working Class Jobs* (New York: Columbia University Press, 1981); Dick Hebdige, *Subculture: The Meaning of Style* (New York: Routledge, 1979).

31. David Morley, *The "Nationwide" Audience: Structure and Decoding* (London: British Film Institute, 1980).

32. During, Introduction, 6.

33. Stuart Hall, "Encoding/Decoding," in *Culture, Media, Language: Working Papers in Cultural Studies, 1972–79,* ed. Stuart Hall, Dorothy Hobson, Andrew Lowe, and Paul Willis (New York: Routledge, 1980), 117–27.

34. Hall, 117.

35. Hall, 124.

36. In fact, During understands these two to be the two primary commitments across the otherwise heterogeneous projects that fall under the penumbra of cultural studies. See During, Introduction.

37. Sheryl Hamilton, "Considering Critical Communication Studies in Canada," in *Mediascapes: New Patterns in Canadian Communication*, ed. Leslie R. Shade (Scarborough, Ont.: Nelson Education) 4–24.

38. During, Introduction.

39. There are notable exceptions, of course, for example, Jennifer Slack and Greg Wise's argument for communication as a way of seeing technology in culture. Slack and Wise, *Culture + Technology: A Primer* (New York: Peter Lang, 2005).

40. Geoffrey Winthrop-Young, "Cultural Studies and German Media Theory," in *New Cultural Studies: Adventures in Theory,* ed. Gary Hall and Claire Birchall (Edinburgh: Edinburgh University Press, 2006), 88–104; Geoffrey Winthrop-Young, "The Kultur of Cultural Techniques: Conceptual Inertia and the Parasitic Materialities of Ontologization," *Cultural Politics* 10, no. 3 (2014): 376–88.

41. Winthrop-Young, "Cultural Studies and German Media Theory," 91.

42. Winthrop-Young, "Cultural Studies and German Media Theory" and "The Kultur of Cultural Techniques."

43. Friedrich Kittler, *Optical Media: Berlin Lectures, 1999,* trans. Anthony Enns (Malden, Mass.: Polity, 2010), 163.

44. Friedrich Kittler, *Gramophone, Film, Typewriter*, trans. Geoffrey Winthrop-Young and Michael Wutz (Stanford: Stanford University Press, 1999).

45. Kittler, *Optical Media.*

46. Kittler, *Gramophone, Film, Typewriter,* 34, 153.

47. Kittler, 157.

48. Friedrich Kittler, *Discourse Networks 1800/1900,* trans. Michael Metteer and Chris Cullens (Stanford: Stanford University Press, 1990), 369.

49. Kittler, 369.

50. David Wellbery, foreword to Kittler, *Discourse Networks*, xii.

51. Kittler, *Optical Media*, 36.

52. Kittler, *Gramophone, Film, Typewriter.*

53. Kittler, *Discourse Networks*, 369.

54. Jussi Parikka, *What Is Media Archaeology?* (New York: Polity, 2012), 68.

55. Erkki Huhtamo and Jussi Parikka, "Introduction: An Archaeology of Media Archaeology," in *Media Archaeology: Approaches, Applications, and Implications*, ed. Erkki Huhtamo and Jussi Parikka (Berkeley: University of California Press, 2011), 9.

56. Wellbery, Foreword.

57. John Durham Peters, Introduction to Kittler, *Optical Media*, 1–18.

58. James Carey, *Communication as Culture, Revised Edition: Essays on Media and Society (New edition)* (New York: Routledge, 1989); Claude E. Shannon, "A Mathematical Theory of Communication," *Bell System Technical Journal* 27 (July 1948): 379–423.

59. Peters, Introduction, 7.

60. See Harold Innis, *Empire and Communications* (Toronto: Dundurn, 2007); Harold Innis, *The Bias of Communication*, 2nd ed. (Toronto: University of Toronto Press, Scholarly Publishing Division, 2008); Marshall McLuhan, *Understanding Media: The Extensions of Man*, ed. W. Terrence Gordon (Corte Madera, Calif.: Gingko Press, 2003).

61. Kittler, *Optical Media*, 31.

62. Sybille Krämer, "The Cultural Techniques of Time Axis Manipulation: On Friedrich Kittler's Conception of Media," *Theory, Culture & Society* 23, nos. 7–8 (2006): 98.

63. Nerone, "Approaches to Media History."

64. Peters, Introduction, 5.

65. Peters points out this tired joke, which is not only telling of Kittler's sense of humor but more prominently of his elitist political convictions.

66. Michel Foucault, *"Society Must Be Defended": Lectures at the Collège de France, 1975–76*, trans. David Macey (New York: Picador, 2003), 10–11.

67. Huhtamo and Parikka, Introduction, 6.

68. Wolfgang Ernst, *Digital Memory and the Archive*, ed. Jussi Parikka (Minneapolis: University of Minnesota Press, 2013); Wolfgang Ernst, *Stirrings in the Archives: Order from Disorder*, trans. A. Siegel (Lanham, Md.: Rowman and Littlefield, 2015); Kittler, *Discourse Networks*; Friedrich Kittler, *Literature, Media, Information Systems*, ed. John Johnston (New York: Routledge, 1997); Kittler, *Gramophone, Film, Typewriter*; Kittler, *Optical Media*; Parikka, *What Is Media Archaeology?*; Bernhard Siegert, *Relays: Literature as an Epoch of the Postal System* (Stanford: Stanford University Press, 1999); Bernhard Siegert, *Cultural*

Techniques: Grids, Filters, Doors, and Other Articulations of the Real, trans. Geoffrey Winthrop-Young (New York: Fordham University Press, 2015); Geoffrey Winthrop-Young, "Drill and Distraction in the Yellow Submarine: On the Dominance of War in Friedrich Kittler's Media Theory," *Critical Inquiry* 28, no. 4 (2002): 825–54; Geoffrey Winthrop-Young, *Kittler and the Media* (New York: Polity, 2011); Siegfried Zielinski, *Deep Time of the Media: Toward an Archaeology of Hearing and Seeing by Technical Means,* trans. Gloria Custance (Cambridge: MIT Press, 2006); Siegfried Zielinski, *[. . . After the Media] News from the Slow-Fading Twentieth Century,* trans. Gloria Custance (Minneapolis: Univocal, 2013).

69. Parikka, *What Is Media Archaeology?,* 6.

70. Parikka, 6.

71. Jussi Parikka, *Insect Media: An Archaeology of Animals and Technology* (Minneapolis: University of Minnesota Press, 2010); Jussi Parikka, "Insects and Canaries: Medianatures and Aesthetics of the Invisible," *Agelaki* 18, no. 1 (2013): 107–19; Jussi Parikka, *The Anthrobscene* (Minneapolis: University of Minnesota Press, 2014); Jussi Parikka, *A Geology of Media* (Minneapolis: University of Minnesota Press, 2015).

72. Parikka, *A Geology of Media,* 4.

73. Parikka, 139.

74. Parikka, *Insect Media,* xiv.

75. Parikka, *A Geology of Media,* 151–52.

76. John Durham Peters, "Strange Sympathies: Horizons of German and American Media Theory," *Media and Society* 15 (2007): 131–52.

77. Bernard Geoghegan, "After Kittler: On the Cultural Techniques of Recent German Media Theory," *Theory, Culture & Society* 30, no. 6 (2013): 70, 77.

78. Winthrop-Young, "The Kultur of Cultural Techniques," 381–382.

79. Winthrop-Young, 381–82.

80. Winthrop-Young, 381.

81. Geoghegan, "After Kittler," 67.

82. Marcel Mauss, "Techniques of the Body," *Economy and Society* 2, no. 1 (1973): 70–88.

83. Geoghegan, "After Kittler," 67.

84. Thomas Macho, "Second-Order Animals: Cultural Techniques of Identity and Identification," trans. Michael Wutz, *Theory, Culture & Society* 30, no. 6 (2013): 30–47.

85. Bernhard Siegert, "The Map Is the Territory," *Radical Philosophy* 169 (Sept./Oct. 2011): 15.

86. Winthrop-Young, "The Kultur of Cultural Techniques," 385.

87. Winthrop-Young, 385.

88. Winthrop-Young, 386–87.

89. Siegert, *Cultural Techniques.*

90. Winthrop-Young, "The Kultur of Cultural Techniques," 387.

91. Siegert, *Relays.*

92. Bernhard Siegert, "Cultural Techniques: Or the End of the Intellectual Postwar Era in German Media Theory," *Theory, Culture & Society* 30, no. 6 (2013): 48.

93. Siegert, 57.

94. Siegert, 58–59.

95. Bernhard Siegert, "Cacography or Communication? Cultural Techniques in German Media Studies," *Grey Room* 29 (2007): 31.

96. Geoghegan, "After Kittler," 79.

97. Winthrop-Young, "The Kultur of Cultural Techniques," 387.

98. Bruno Latour, *Reassembling the Social: An Introduction to Actor-Network-Theory* (Oxford: Oxford University Press, 2005), 86n106.

99. Rhetoric of science, now rhetoric of science, technology, and medicine (RSTM), was not typically included under the STS umbrella as early as the other fields listed here. We agree with rhetorician Scott Graham and others that, viewed historically and even more pointedly in recent years, RSTM is a field that informs and is informed by STS. One simple and marked piece of evidence for this argument is the uptake of Latourian flat ontologies/actor-network theory to address some of the very same problematic binaries between the material and the socio-cultural that we describe throughout this chapter. S. Scott Graham, *The Politics of Pain Medicine: A Rhetorical-Ontological Inquiry* (Chicago: University of Chicago Press, 2015).

100. Joseph Rouse, "What Are Cultural Studies of Scientific Knowledge?," *Configurations* 1, no. 1 (1988): 57–94.

101. Donna Haraway, "Situated Knowledges: The Science Question in Feminism and the Privilege of Partial Perspective," *Feminist Studies* 14, no. 3 (1988): 575–99; Lorraine Daston and Peter Galison, *Objectivity* (Cambridge, Mass.: Zone Books, 2006).

102. Jeremy Packer and Peter Galison, "Abstract Materialism: Peter Galison Discusses Foucault, Kittler, and the History of Science and Technology," *International Journal of Communication* 10 (2016): 3160–73.

103. Peter Galison, "Material Culture, Theoretical Culture, and Delocalization," in *Science in the Twentieth Century,* ed. John Krig and Dominique Pestre (London: Harwood, 1997), 669–82.

104. Foucault, *The Archaeology of Knowledge.*

105. Jonathan Sterne, *The Audible Past: Cultural Origins of Sound Reproduction* (Durham, N.C.: Duke University Press, 2003), 127–28.

106. Peter Galison, *Einstein's Clocks, Poincaré's Maps: Empires of Time* (New York: Norton, 2003), 47.

107. Peter Galison, "The Collective Author," in *Scientific Authorship: Credit and*

Intellectual Property in Science, ed. Mario Biagoli and Peter Galison (New York: Routledge, 2002), 325–58.

108. Daston Galison, *Objectivity*, 369.

109. Lorraine Daston and Elizabeth Lunbec, *Histories of Scientific Observation* (Chicago: University of Chicago Press, 2011), 1.

110. Lorraine Daston, "Science Studies and the History of Science," *Critical Inquiry* 35, no. 4 (2009): 800. Note that this argument was prefigured by Bruno Latour in "Why Has Critique Run Out of Steam? From Matters of Fact to Matters of Concern," *Critical Inquiry* 30, no. 2 (2004): 225–48.

111. Daston, "Science Studies," 809.

112. Kittler, *Discourse Networks*.

113. Jeremy Packer, "What Is an Archive? An Apparatus Model for Communications and Media History," *Communication Review* 13 (2010): 88–104; Jeremy Packer, "The Conditions of Media's Possibility: A Foucauldian Approach to Media History," in *Media History and the Foundations of Media Studies*, ed. John Nerone (New York: Blackwell, 2013), 88–121.

114. Stuart Hall, *Cultural Studies 1983* (Durham, N.C.: Duke University Press, 2016).

Index

Page references in italic refer to illustrations

KATE MADDALENA is assistant professor in the Institute for Communication, Culture, Information, and Technology at the University of Toronto.

ALEXANDER MONEA is assistant professor in the English department and cultural studies program at George Mason University. He is author of *The Digital Closet: How the Internet Became Straight* and coeditor of *Amazon: At the Intersection of Culture and Capital* (with Paul Smith and Maillim Santiago).

PAULA NUÑEZ DE VILLAVICENCIO is a PhD candidate at the University of Toronto.

KATHLEEN OSWALD is adjunct faculty in the Department of Communication at Villanova University, where she teaches visual communication and new media.

JEREMY PACKER is professor in the Institute for Communication, Culture, Information, and Technology and the Faculty of Information at the University of Toronto. He is coauthor of *Killer Apps: War, Media, Machine* (with Joshua Reeves) and coeditor of *Communication Matters: Materialist Approaches to Media, Mobility, and Networks.*

JOSHUA REEVES is associate professor in the School of Communication at Oregon State University. He is coauthor of *Killer Apps: War, Media, Machine* (with Jeremy Packer) and author of *Citizen Spies: The Long Rise of America's Surveillance Society.*